SpringerBriefs in Complexity

Series Editors

Henry D. I. Abarbanel, Institute for Nonlinear Science, University of California, La Jolla, CA, USA

Dan Braha, New England Complex Systems Institute, University of Massachusetts, North Dartmouth, MA, USA

Péter Érdi, Center for Complex Systems Studies, Kalamazoo College, Department of Physics, Kalamazoo, MI, USA

Karl J Friston, Institute of Cognitive Neuroscience, University College London, London, UK

Hermann Haken, Center of Synergetics, University of Stuttgart, Stuttgart, Germany

Viktor Jirsa, Centre National de la Recherche Scientifique (CNRS), Université de la Méditerranée, Marseille, France

Janusz Kacprzyk, Systems Research Institute, Polish Academy of Sciences, Warsaw, Poland

Kunihiko Kaneko, Research Center for Complex Systems Biology, The University of Tokyo, Tokyo, Japan

Scott Kelso, Center for Complex Systems and Brain Sciences, Florida Atlantic University, Boca Raton, FL, USA

Markus Kirkilionis, Mathematics Institute and Centre for Complex Systems, University of Warwick, Coventry, UK ·

Jürgen Kurths, Nonlinear Dynamics Group, University of Potsdam, Potsdam, Brandenburg, Germany

Ronaldo Menezes, Department of Computer Science, University of Exeter, Exeter, UK

Andrzej Nowak, Department of Psychology, Warsaw University, Warszawa, Poland

Hassan Qudrat-Ullah, School of Administrative Studies, York University, Toronto, Canada

Peter Schuster, University of Vienna, Vienna, Austria

Frank Schweitzer, System Design, ETH Zurich, Zürich, Switzerland

Didier Sornette, Entrepreneurial Risk, ETH Zurich, Zürich, Switzerland

Stefan Thurner, Section for Science of Complex System, Medical University of Vienna, Vienna, Austria

Linda Reichl, Center for Complex Quantum Systems, University of Texas, Austin, TX, USA

SpringerBriefs in Complexity are a series of slim high-quality publications encompassing the entire spectrum of complex systems science and technology. Featuring compact volumes of 50 to 125 pages (approximately 20,000-45,000 words), Briefs are shorter than a conventional book but longer than a journal article. Thus Briefs serve as timely, concise tools for students, researchers, and professionals.

Typical texts for publication might include:

- A snapshot review of the current state of a hot or emerging field
- A concise introduction to core concepts that students must understand in order to make independent contributions
- An extended research report giving more details and discussion than is possible in a conventional journal article,
- A manual describing underlying principles and best practices for an experimental or computational technique
- An essay exploring new ideas broader topics such as science and society

Briefs allow authors to present their ideas and readers to absorb them with minimal time investment. Briefs are published as part of Springer's eBook collection, with millions of users worldwide. In addition, Briefs are available, just like books, for individual print and electronic purchase. Briefs are characterized by fast, global electronic dissemination, straightforward publishing agreements, easy-to-use manuscript preparation and formatting guidelines, and expedited production schedules. We aim for publication 8-12 weeks after acceptance.

SpringerBriefs in Complexity are an integral part of the Springer Complexity publishing program. Proposals should be sent to the responsible Springer editors or to a member of the Springer Complexity editorial and program advisory board (springer.com/complexity).

More information about this series at http://www.springer.com/series/8907

Hassan Qudrat-Ullah

Improving Human Performance in Dynamic Tasks

Applications in Management and Industry

 Springer

Hassan Qudrat-Ullah
School of Administrative Studies
York University
Toronto, Canada

ISSN 2191-5326   ISSN 2191-5334 (electronic)
SpringerBriefs in Complexity
ISBN 978-3-030-28165-6   ISBN 978-3-030-28166-3 (eBook)
https://doi.org/10.1007/978-3-030-28166-3

This Springer imprint is published by the registered company Springer Nature Switzerland AG
The registered company address is: Gewerbestrasse 11, 6330 Cham, Switzerland

To My Mother
Fazeelat Begum, who despite having limited
resources and no formal education was able
to provide the best possible education and
training to us (five brothers and one sister)
with hope-based attitude and inspirational
stance on several conflicting and difficult
situations. Mom, I still owe you a lot and still
need your prayers.
-Hassan Qudrat-Ullah-

Preface

Improving human performance in complex, dynamic tasks has always been at the forefront of both research and practice of organizational decision making. Simulation-based education and training is a multibillion-dollar industry. The purpose of this book is to provide the reader with the knowledge about the design, development, validation, and application of an innovative, system dynamics-based interactive learning environment that includes a systematic debriefing. Specifically, a laboratory experiment is reported in which the participants managed a dynamic task by playing the roles of fishing fleet managers. A comprehensive model consisting of five evaluation criteria, (i) task performance, (ii) decision strategy, (iii) decision time, (iv) structural knowledge, and (v) heuristics knowledge, is developed and used. The key insights gleaned from the empirical data include the following: (i) the process-oriented debriefing improves subjects' performance better than the outcome-oriented debriefing, and (ii) contrary to the cost-benefit approach to decision making, more systematic effort is needed to perform better in dynamic tasks.

In the quest for innovative solutions for the education and training of people in dynamic tasks, many challenges lie ahead. Specifically, as we move towards displacing traditional thinking that people perform poorly in dynamic tasks, founded in dominant dynamic decision making literature, to one where plural logics of "systematic debriefing-based training with SDILEs" coexist under conditions of uncertainty and ambiguity, the need for systematic and integrated solutions for improving human performance in dynamic tasks becomes pronounced. Our aim here has been to focus our attention on the whole virtuous cycle of expertise development: decision making $\rightarrow$ learning $\rightarrow$ decision making. It is our hope that this book will stimulate a new way of thinking as a proclamation of a new era of resource constraints and a renewed focus on "integrative" solutions for people's education and training in dynamic tasks.

Dhahran, Saudi Arabia   Hassan Qudrat-Ullah

Acknowledgments

I would like to thank everyone, mentioned herein or not, for their continued support in helping to bring this book to completion. My appreciation also goes to all the people at Springer, USA, especially Christoph Baumann and Jeffrey Taub with whom I corresponded for their advice and facilitation in the production of this book. I would like to thank Chandhini Kuppusamy and the production team from Springer for their help in the final production of this book.

I am grateful to my family, Tahira Qudrat, Abdul Jabbar, Anam Qudrat, Ali H. Qudrat, Umer K. Qudrat, Umael H. Qudrat, and Abdur-Rahman Hassan, for their encouragement and support throughout this endeavor. It would be unfair not to mention Saleem (all doer), Naheem (the spender), Naveed (the talent in progress), Wasim (still the youngest), and Zahida (the ultimate patience practitioner) for their prayers and support. Special thanks and appreciation go to my mother-in-law, Saira Bano, and father-in-law, Allah Ditta, whose prayers and wishes provide unique strengths to me for such taxing tasks.

Finally, the author would like to acknowledge the financial support provided by the Deanship of Scientific Research (DSR) at King Fahd University of Petroleum and Minerals (KFUPM) for funding this work through project No. BW181001.

Hassan Qudrat-Ullah
Dammam, Saudi Arabia
July 2019

Acknowledgments

Content Overview

The integrating theme of this book is system dynamics-based perspective, cognitive apprentice approach, and learning principles to improve human performance in dynamic tasks. The book contains six chapters. Chapter 1, "Decision Making and Learning in Dynamic Tasks," presents the concept of "dynamic task" and explains why learning and decision making in dynamic tasks are hard, what are the key challenges to decision making and learning in dynamic tasks, why system dynamics-based interactive learning environments (SDILEs) are an effective tool to improve people's decision making in dynamic tasks, and why is the incorporation of debriefing into the design of an SDILE critical. Chapter 2, "SDILEs in Service of Dynamic Decision Making," provides an overview and elaborates on the implementation of (i) HCI design principles, (ii) cognitive apprenticeship theory and *Gagné*'s nine instructional events, and (iii) structured debriefing and learning-inducing elements of any SDILE with the example of our developed and validated SDILE, SIADH-ILE. Also, a five-dimensional evaluative model is presented. Chapter 3, "The Experimental Approach," explains the experimental approach through various dimensions. First, the research design is elaborated. Second, the dynamic task, SIADH-ILE, its casual structure, mathematics model, and interface design are explained. Third, the protocol of the experiment is also described. Chapter 4," Results of Experimental Research," presents our key results. Both the main effects and indirect effects of debriefing on these five dimensions are reported. Beginning with the descriptive analysis of our participant, we report the subjects' performance and learning in the dynamic task. Subjects' reaction to received debriefing as well as the effects of the practice is also reported here. Chapter 5, "Discussion and Conclusions," presents key limitations of this study, our major findings, and implications of dynamic decision making research and of improving practice in dynamic tasks in various domains including computer simulation-based education and training, aviation, healthcare, and policymaking. We will also talk about how the users perceived the utility of SIADH-ILE in improving their decision making and learning in the dynamic task.

Finally, Chap. 6, "Future Research Directions in Dynamic Decision Making," highlights the areas of research for future researchers.

Contents

Chapter 1
Decision Making and Learning in Dynamic Tasks

Abstract Successful decision making is a necessary and sufficient condition for effective and efficient management of any business or organization. In this chapter, we introduce the concept of "dynamic task" and explain why learning and decision making in dynamic tasks is hard, what are the key challenges to decision making and learning in dynamic tasks, and why system dynamics-based interactive learning environments (SDILEs) are an effective tool to improve people's (In this book the words, people, users, learners, and decision makers are used interchangeably.) decision making in dynamic tasks. Why is the incorporation of debriefing into the design of an SDILE critical? Through such questions and assertions, the objective of this introductory chapter is to entice the reader for the following material in this book.

Keywords Aviation · Causal understanding · Cognitive ability · Correlational heuristics · Decisional aid · Dynamic complexity · Debriefing · Dynamic task · Efficacy of SDILEs · Feedback loops · System Dynamics

1.1 Introduction

We make decisions primarily to make efficient and effective use of our limited resources. All organizations have limited resources. It is incumbent upon them to have better decisions and have better decision makers onboard all the times. Businesses and organizations spend a good amount of money and time to educate and train their employees. The nature of most of the organizational decisions is strategic and dynamic. In dynamic tasks which are characterized by having time lags (e.g., between an action and its consequence), multiple and interdependent decisions, nonlinear relationships between the variables of the task system, and feedback loops add to the difficulty of managerial decision making. Those who are managing the tasks in aviation (e.g., flying planes), emergency room operations (e.g., neurosurgery), and policy domain (e.g., design and assessment of health, education, energy policies) know at first hand why it is hard to make the decision in these tasks. Despite this difficulty, people are successfully doing these dynamic tasks. Both formal education and on-the-job training play a critical role in making these individual good decision makers.

In today's fast-paced, highly competitive, and technology-intensive environment, organizations hardly can afford to send their people back to school for formal education and training. Instead, workshop-based short training programs that allow people to better understand the dynamic tasks are high in demand. System dynamics methodology (Forrestor 1961; Sterman 2000) with its fundamental premise, "structure (of any task system) drives (its) performance," provides a viable solution to this managerial need. Management flight simulators and interactive learning environments are built using a system dynamics model (SDILEs) to train people in dynamic tasks. Despite the widespread use of these SDILEs, their success in improving people's decision making in dynamic tasks is mixed at best. Researchers (Sterman 2000; Spector 2000; Größler et al. (2016) have suggested that any mechanism that helps the decision makers to better understand the structure of the task system should be incorporated into the design of any SDILE. Building on some learning principles and theories together with the principles of design science, therefore, here in this book, our purpose is to introduce promising alternative, debriefing-based SDILEs, SIADH-ILE, as a viable decisional aid for the education and training of people in dynamic tasks. We will empirically test SIADH-ILE with a unique five-dimensional evaluation model and then present it to you, the readers of this book. We hope this book will facilitate you to embrace and adapt debriefing-based SDILE to the training program of your organization or embark on a new SDILE-based workshop-oriented solution for the training needs of your people.

1.2   What Is a Dynamic Task?

We all make decisions be it in a personal situation or in a business context. In the context of managerial decision making, a raison d'etre for modern-day managers, most of the decisions are dynamic in nature. In contrast to static decision making situations where decisions are often one-shot (e.g., locating a destination on a map, finding the shortest distance between two locations (given the fixed existing routes and route conditions), and playing a lottery are all examples of static decision making), dynamic tasks are different in many respects. In dynamic tasks not only (i) multiple decisions are required, but (ii) these multiple decisions are interdependent, (iii) decision making situation changes due to the decision made or independent of the decisions made or both, and (iv) these decisions involve feedback loops (Edwards 1962; Qudrat-Ullah 2010). Consider the example of driving a car. On a sunny day, you decide to take your family to an adventurous park involving a 2-hour ride from home. Your driving involves driving on various roads (local street, main roads, and highways) and zones (e.g., a school zone). Your task of driving, from home to the park, is a dynamic task. You can't live with just a single decision (e.g., set the cruise and reach the destination). Instead, a series of decisions are needed (e.g., have to observe different speeds due to changing zones (e.g., schools and hospital zones require different speed), speed limits (e.g., local roads and highways have different speed limits), and road conditions (e.g., slow speed is required at turns)). Besides these speeding decisions, you have to make several other decisions including monitoring your car's conditions (e.g., temperature, fuel), observing the flow of traffic closely, and paying attention to weather conditions. Secondly, driving decisions are interdependent, i.e., the outcome of a prior decision, say made at time t_{n-1}, has an influence on the current decision, t_n (e.g., your current speeding decision on a smooth and straight road depends on the successful prior decision made at turns). An unsuccessful decision made at turns (e.g., see Fig. 1.1) can simply jeopardize the successful completion of the task itself.

Thirdly, situations change (e.g., internally due to prior poor decisions made at turns or externally due to changing weather conditions). Finally, driving decisions involve feedback loops. For instance, you make a decision to reach a target speed of your car, you observe the reading on the speedometer (which feeds back to you with the new information about the current speed of your car, and if you ignore this feedback, then you might face the undesirable consequence (e.g., an accident)), and then you readjust the speeding decision, a feedback loop, as shown in Fig. 1.2. It can be noted that this decision making during driving can serve as both personal decision making and business or managerial decision making situation.

Generally, in the field of managerial decision making, there are two main categories of decisions: (i) operational decisions which are short-term, focused on day-to-day activities of the firm (e.g., manpower scheduling), and (ii) strategic decisions which are long-term, strategies to achieve the strategic goals of the organization (e.g., capacity expansion and outsourcing decisions). One could argue that the nature of "driving decision making" falls into the category of operational decisions. What about strategic decisions, a critical category of decisions for the majority of

Fig. 1.1 Driving a car is a dynamic task. (Image source: http://exchange.aaa.com/automobiles-travel)

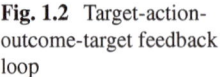

Fig. 1.2 Target-action-outcome-target feedback loop

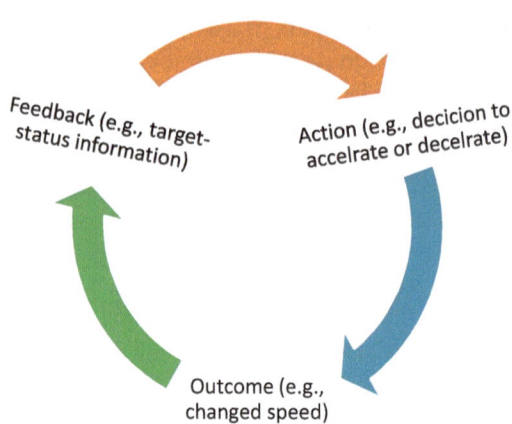

Feedback (e.g., target-status information)

Action (e.g., decicion to accelrate or decelrate)

Outcome (e.g., changed speed)

business firms' sustainable growth and prosperity. For instance, the human resource management department of a business firm deals with "hiring decisions," which are the critical strategic decisions of the firm.

On the surface, perhaps hiring decisions involve well-defined tasks such as job advertisement, screening of applicants, and testing/interviewing of the applicant, and finally, the offer is extended to the selected applicant. However, to consider a hiring decision as such is too simplistic. Consider the example of a new hire who does not perform as per the expectation. The manager of this employee often will send this employee for counseling and training. Now after additional training, this employee is still not at par with the rest of his colleagues. Depending on the rules and regulation of the company as well as of the host country where this company is operating, the manager can either fire the employee or keep him with a hope that he will improve his performance. In either case, the company has to incur the additional cost. For instance, in the case the low-performing employee is fired, its replacement cost is imminent. In some cases, costly litigation might be faced by the company. On the other hand, if the manager keeps this low-performing employee, then such an employee will soon seek "buddies" around him. As a result, more employees are becoming less productive. When more and more employees become less productive, the company suffers: even sometimes bankruptcy is the only option left for the firm. This hypothetical yet highly probable hiring decision situation shows that hiring decisions are both strategic and dynamic in nature. Other real-life decisions such as controlling the money supply, achieving sustainable use of resources, flying a plane, doing an ER operation, management of healthcare systems, design of sustainable energy policies of a region or a country, and management of global supply chains are all dynamic tasks.

Also, time-critical tasks such as traders in financial markets, pilots in combat situations, doctors in emergency rooms, air-traffic controllers, managers in complex manufacturing settings, commanders in battle situations are all dynamic decision making situations: decision makers are rushed to make decisions in rapidly changing conditions. Thus, dynamic tasks involve (i) a series of multiple and interdependent decisions, (ii) changing decision environment(s), and (iii) multiple feedback loops with embedded time delays and uncertainties. Next, we discuss people performance in dynamic tasks and how they can be improved.

1.3   Learning and Decision Making in Dynamic Tasks

Learning and decision making in dynamic tasks is difficult at best. Consider the well-known phenomenon of Friendly Fires: An event when in a military operation, the force fires, mistakenly, on its own troops. Such events are also known as "fratricides." "A fratricide is the employment of friendly weapons and munitions with the intent to kill the enemy or destroy his equipment or facilities, which results in unforeseen and unintentional death or injury to friendly personnel (USDOA 1993, p. 1)."

Despite state-of-the-art technologies and training programs at the disposal of our decision makers, fratricide rates are increasing. In a highly stressful and fast-paced environment, the decision makers have to take action without having the luxury of any testing and learning prior to actual decision making. Thus, errors happen, leading to poor task performance. Even in relatively less stressful environments like the control rooms of nuclear power plants or the cockpit of airplanes, which are built with highly ergonomic guidelines, most of the accidents happen due to human failures. This, we continue to face the fundamental question: how to improve people's decision making in dynamic tasks? Our best hope is to educate and train them.

However, when it comes to decision making and learning in dynamic tasks, there are several impeding factors (Sterman 2000; Qudrat-Ullah 2010).

1. *Understanding dynamic complexity is hard.* In dynamic tasks, actions and consequences are hardly related in both time and space. Our ability to understand the effects of time lags between our actions and their consequences coupled with the interactions between feedback loops (e.g., as shown in Fig. 1.2) is very limited.
2. *Availability and quality of information are limited.* Human beings are notorious in overestimation or underestimation of information. Our communication is often plagued with oversimplification, distortion, delays, and biases.
3. *Cognitive ability to process information is limited.* Human cognition faces difficulty in processing information when faced with causally related information. Often people adopt an event-based, open-loop view of causality, ignore feedback processes, fail to appreciate time delays and are insensitive to nonlinearities present in the feedback loop structures of the task system, perceive flawed cognitive maps of the causal structure of the systems, make erroneous inferences even about the simplest possible feedback systems, and fall prey to judgmental errors and biases and defensive routines (Sterman 1994, 1989).
4. *Use of correlational heuristics abounds.* In dynamic decision making (DDM) literature, time and again it is reported that people use correlation heuristics instead of a structure-behavior understanding of the task variables.

Any training aid or decision support system for decision makers, therefore, should allow the users to overcome such impediments to decision making and learning in dynamic tasks.

1.4   Interactive Learning Environments (ILEs) and Dynamic Decision Making (DDM)

Learning about several tasks often involves some kind of prototyping or pilot testing or experimentation. However, in the case of many real-world dynamic tasks (e.g., medical emergency treatment skills, nuclear power accident management training, and training of new pilots), real-world experimentation is hardly a possibility. To better prepare people for such dynamic tasks, the education and training need has long motivated the design, development, and application of computer simulation-based interactive learning environments (ILEs) as a decision support system (Lane 1995; Sterman 2000). In particular, with the impressive advancements in technology and developments in system dynamics field, the development and use of system dynamics-based ILEs (SDILEs) are on the rise. An SDILE comprises of (i) an underlying system dynamics model of the dynamic task, (ii) an interactive user interface, (iii) a user/learner/decision maker, and (iv) the human facilitation or support. In addition, an ILE has to have a clear learning objective. Therefore, in our conception of an ILE, we use "ILE" as a term sufficiently general to include microworlds, management flight simulators, simulation-based learning laboratories, and any other computer simulation-based environment—the domain of these terms is all forms of action whose general goal is the facilitation of learning (Qudrat-Ullah 2014; Spector 2000). The gameplay is commonly characterized as a voluntary, nonproductive task separated from the real world, while learning in an ILE is usually nonvoluntary, aiming at specific learning outcomes, and related to the real life (Garris et al. 2002). Therefore, most of the video games or simulations which are played "just for fun" will not be included in this definition of an ILE. An SDILE is considered an effective decisional aid for improving users' performance in the dynamic task due to its ability account for the feedback structures of a dynamic task. Training with SDILEs is often delivered in a workshop setting (Qudrat-Ullah 2010, 2014). Some well-known SDILEs are People Express, Beefeater, Strategem-2, the Fishbanks game, B and B Enterprise, Learn!, Market Growth, C-learn, Friday Night at ER, SIADH-ILE, and FishBankILE (Qudrat-Ullah et al. 1997).

Despite their widespread use for people's training for decision making in dynamic tasks, the effectiveness of SDILEs is mixed at best. For instance, a large body of empirical research indicates that during sessions with SDILEs, people perform poorly on dynamic tasks (Diehl and Sterman 1995; Fischer and Gonzalez 2016; Qudrat-Ullah 2010; Sterman 1994, 2000; Sterman and Dogan 2015; Sterman and Sweeney 2007). Key findings from this empirical literature regarding why people perform poorly in dynamic tasks are:

(i) They do not develop adequate models of the task system. Instead, they continue to have a simplified linear view of the task system.

(ii) They lack causal understanding of the task system and often apply correlational heuristics (e.g., they rarely recognize and appreciate that there exists a time lag between decisions and their effects; the relationships between various variables of the task system are nonlinear).

(iii) They lack a global view of the task system. Instead, they focus on the local details of the task system.

(iv) They misperceive the feedback orientation of the dynamic tasks.

(v) Learned knowledge from SDILEs is not well assimilated with the existing knowledge of the users.

It is apparent that for any effective learning environment, the mere availability of simulation will not address these issues. Instead, the design of an SDILE should be based on sound human-computer interaction (HCI) design elements and relevant learning principles and theories. Therefore, departing from the focus of DDM research of the past two decades on reporting people's poor performance in dynamic tasks, this book takes a solution-oriented view and presents with some viable alternate design of SDILEs to promote user decision making and learning in dynamic tasks. In the next chapter, we present some relevant HCI design elements and learning principles for the design of SDILEs and how we incorporated these into the design of our SDILE, SIADH-ILE.

1.5 Summary

Given the unprecedented nature of competition, a globalized world, and the advancement of technology, improving decision making and learning of people in dynamic tasks has become a much sought after need of today's organizations. SDILEs provide a viable solution. To improve the efficacy of SDILEs in enhancing people's decision making and learning in dynamic tasks, it is argued that the design of an SDILE should incorporate both HCI design guidelines and learning principles.

References

Diehl, E., & Sterman, J. D. (1995). Effects of feedback complexity on dynamic decision making. *Organizational Behavior and Human Decision Processes, 62*(2), 198–215.

Edwards, W. (1962). Dynamic decision theory and probabilistic information processing. *Human Factors, 4*, 59–73.

Fischer, H., & Gonzalez, C. (2016). Making sense of dynamic systems: How our understanding of stocks and flows depends on a global perspective. *Cognitive Science, 40*(2), 496–512.

Forrester, J. W. (1961). *Industrial Dynamics*. Cambridge, MA: Productivity Press.

Garris, R., Ahlers, R., & Driskell, J. E. (2002). Games, motivation, and learning: A research and practice model. *Simulation & Gaming, 33*(4), 441–467.

Größler, A., Rouwette, E., & Vennix, J. (2016). Non-conscious vs. deliberate dynamic decision-making—A pilot experiment. *Systems, 4*(13), 1–13. https://doi.org/10.3390/systems4010013.

Lane, D. C. (1995). On a resurgence of management simulations and games. *Journal of the Operational Research Society, 46*, 604–625.

Qudrat-Ullah, H. (2010). Perceptions of the effectiveness of system dynamics-based interactive learning environments: An empirical study. *Computers and Education, 55*, 1277–1286.

Qudrat-Ullah, H. (2014). Yes we can: Improving performance in dynamic tasks. *Decision Support Systems, 61*, 23–33.

Qudrat-Ullah, H., Saleh, M. M., & Bahaa, E. A. (1997). Fish Bank ILE: An interactive learning laboratory to improve understanding of 'The Tragedy of Commons'; a common behavior of complex dynamic systems. *Proceedings of 15th international system dynamics conference*, Istanbul, Turkey.

Spector, J. M. (2000). System dynamics and interactive learning environments: Lessons learned and implications for the future. *Simulation and Gaming, 31*(4), 528–535.

Sterman, J. D. (1989). Modeling managerial behavior: Misperceptions of feedback in a dynamic decision making experiment. *Management Science, 35*, 321–339.

Sterman, J. D. (1994). Learning in and about complex systems. *System Dynamics Review, 10*, 291–330.

Sterman, J. D. (2000). *Business dynamics: Systems thinking and modeling for a complex world.* New York: McGraw-Hill.

Sterman, J. D., & Dogan, G. (2015). I'm not hoarding, I'm just stocking up before the hoarders get there. Behavioral causes of phantom ordering in supply chains. *Journal of Operations Management, 39–40*, 6–22.

Sterman, J. D., & Sweeney, B. (2007). Understanding public complacency about climate change: Adults' mental models of climate change violate conservation of matter. *Climatic Change, 80*(3–4), 213–238.

USDOA. (1993, August 31). *US Department of the Army, Military Operations: U.S. Army Operations Concept for Combat Identification, TRADOC Pam 525-58* (p. 1). Fort Monroe, VA: Training and Doctrine Command.

Chapter 2
SDILEs in Service of Dynamic Decision Making

Abstract As the objective of a SDILE is to improve people's decision making and learning in dynamic tasks, its design should incorporate the mechanisms to support people's learning. Researchers in the SD community have identified three such mechanisms to be an essential part of a SDILE: (i) HCI design principles, (ii) cognitive apprenticeship theory and *Gagné*'s nine instructional events, and (iii) structured debriefing. We provide an overview and elaborate on the implementation of these learning inducing elements of any SDILE with the example of our developed and validated SDILE, SIADH-ILE. Also, to better assess the efficacy of SDILEs and to fully capture the decision makers' performance in dynamic tasks, we present a five-dimensional evaluative model. Based on this newly developed evaluative model, we advance five assertions pertaining to the efficacy of debriefing-based SDILEs.

Keywords Dynamic tasks · Evaluative model · HCI design principles · Instructional events · Cognitive apprenticeship theory · SDILE · Structured debriefing · Situated learning · Modeling and explaining · Facilitator support · SIADH-ILE · Decision strategy · Decision time · Structural knowledge · Heuristics knowledge

2.1 Introduction

Decision making in and about dynamic tasks using SDILEs have received substantial attention of researchers as well as practitioners. Everyone wants to make better decisions and improve the capability of their employees to make a better decision in dynamic tasks. The use of simulations or simulation-based learning environment is not new (Faria 1998; Mayer et al. 2011; Tamara et al. 2013). For instance, training of people in the aviation industry alone is a multibillion dollar industry (NTSA 2011). Simulator-based training is a rich area both for the researchers and practitioner. Users of these training tools want efficient as well as low-cost solutions.

Despite the increasing use of SDILEs both in research and practice about improving people' decision making and learning in dynamic tasks, there is an ongoing urge to improve the efficacy of these learning environments. Researchers have suggested several ways to improve the design of SDILEs. For instance, Howie et al. (2000) confronted the misperceptions of feedback hypothesis—people perform poorly in dynamic tasks because they poorly understand the dynamics of feedback loops of the task system and present an alternate design. He implemented interface in an SDILE. Participants in his experimental study showed improved performance. Use of HCI design is considered as a useful intervention to enhance the efficacy of SDILE.

Learning theories and principles inform us how people actually learn. System dynamics community, in particular, has advocated the use of Gagné's instructional theory and Collins' apprenticeship approach (Gagné 1985, Collins). Likewise, the use of structured debriefing into the design of SDILEs has successfully received the attention of researchers and practitioners of SDILEs. People face difficulties in decision making and learning with any kind of simulation albeit with SDILEs. They face the challenges of (i) assimilating new knowledge, learned in the session with the simulated task, (ii) which aspects of their learning are relevant to real-world decision making, and (iii) they might need clarity about some concepts or relationships among the variables of the task system. To overcome these learning impediments, the use of these learning theories, principles, and mechanisms is critical. In the absence of such support mechanisms specifically the structured debriefing, users of simulations might win the game but can hardly learn about the dynamic tasks. Therefore, the incorporation of these decisional aids or support mechanisms into the design of a SDILE is immensely important. When we reviewed much of prior experimental research pertaining to DDM, we found only a few studies that have utilized these mechanisms. Thus, in the context of dynamic tasks, perhaps this is the first study that has utilized all of these mechanisms simultaneously in the design of a SDILE, SIADH-ILE.

In this chapter, therefore, we review and demonstrate the implementation of (i) HCI design principles, (ii) learning principles, and (iii) structured debriefing into the design of our SDILE, SIADH-ILE. In stark contrast to the prior research in DDM, to effectively measure the performance of decision makers in dynamic tasks and evaluate the efficacy of SDILEs, we present a five-dimensional evaluation model. This evaluative model accounts for subjects' decision making and learning in dynamic tasks through five measures:

 (i) How do they perform the task?
 (ii) Did they use systematic and pattern recognition-oriented decision strategies?
(iii) Did they become efficient decision makers?
(iv) To what extent they developed structural knowledge about the task system?
 (v) Did they develop some useful heuristics?

Finally, we present our five formal hypotheses that we empirically tested in our study and have reported in Chap. 5 in this book. Next, Chap. 4 provides the details about methodology: the laboratory experimental approach.

2.2   Use of HCI and Learning Principles in the Design of SDILEs

The main objective of SIADH-ILE is to support the user's learning and decision making about a dynamic task and managing HIV/AIDS dynamics in Canada. Both the HCI design principles and principles of learning theory were incorporated into the design of SIADH-ILE. Drawing on the empirical findings of Howie et al. (2000), we have utilized these HCI principles as described in Table 2.1.

Learning theories and principles inform the effective design of simulation-based systems like SDILEs. Here we review two relevant theories: (i) Collin's cognitive apprenticeship theory (Collins 1991) and (ii) *Gagné*'s learning theory (Gagné 1985). Insights gains from this literature will inform the design of our SDILE, SIADH-ILE.

2.2.1   Cognitive Apprenticeship

The literature of cognitive apprenticeship theory (Collins 1991) facilitates decision making and learning in complex dynamic environments (Davidsen and Spector 1997). Cognitive apprenticeship theory emphasizes on situated learning, modeling of process knowledge and expert performance to be learned, provision of timely coaching of the new learners, articulation by the learners of their knowledge and strategies applied, forming and testing of hypotheses by the learners through a systematic exploration of the dynamic environment, and reflection on learner's performance both by the learner and the trained instructor/facilitator.

Note the critical dimensions of the facilitator support according to the cognitive apprenticeship model. The facilitator support is crucial in almost all the facets of cognitive apprenticeship-*situated learning, modeling and explaining, coaching, reflection, articulation, and exploration* (as shown in Fig. 2.1).

Situated learning refers to learning knowledge and skills in contexts that reflect the way knowledge will be useful in real life (Collins 1991). In the context of decision making and learning with ILEs, the facilitator sets the context of learning at the

Table 2.1 Incorporation of relevant HCI principles

HCI principle	Implementation mechanisms
Utilize people's prior knowledge	We used metaphors in the *help menu* of SIADH-ILE's interface
Tap into people's pattern recognition capabilities	We developed *history function* in the user interface where users are presented with their performance both in graphs and tables
Make the relationship between task system variables more salient	With the use of the *information system* of SIADH-ILE, users are presented with relevant exemplars

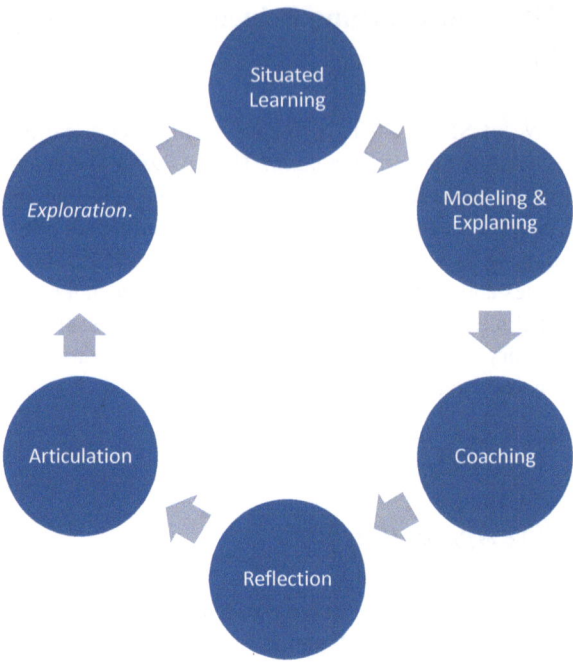

Fig. 2.1 Collins's apprenticeship model

outset of the session, for example, by presenting about the problem domain, issues, and objectives of the session. Given that the context of learning is well designed and presented by the facilitator, learners are more likely to induce transfer learning. That is, after having the rich experiences of decision making and learning with the computer-simulated environment of the SDILE, the knowledge learned by the decision makers is stored in a form that is usable in similar or novel contexts of the real-world problems.

Modeling is showing how a process unfolds, and explaining refers to reasoning about why it happens that way (Collins 1991). In an SDILE, the in-task and post-task level facilitator support can embrace modeling and explaining characteristics of cognitive apprenticeship. In debriefing reviews, a facilitator, with the use of technology, can effectively explain to the decision makers what happened and what did not happen. Through facilitator's debriefings, decision makers can also see the "expert" solution. They are likely to better assimilate the new knowledge they have learned while resolving the dynamic decision task of the ILE.

Coaching enables decision makers to do the task they might not otherwise be able to accomplish (Collins 1991). In SDILEs' sessions, the facilitator's coaching is very important. For example, coaching can arouse decision makers' interest in the decision task, provide help at critical times, and gratify their accomplishments. The coach/facilitator can point out things that do not go as expected and explain why.

Reflection pertains to learners looking back over what they did and analyzing their performance (Collins 1991). As mentioned earlier, debriefing reviews by the facilitator provide excellent support for decision makers' reflection on performance. The facilitator can replay decision makers' performance on decision task as data to be analyzed. A description of various heuristics in terms of the process the decision makers are reflecting on can help the decision makers to expose and correct their "entrenched beliefs" about the phenomena under study. Likewise, the debriefing reviews provide an opportunity for the decision makers to articulate—making their tacit knowledge explicit.

Exploration involves pushing students to test different hypotheses and assumptions to observe their effects (Collins 1991). In an SDILE session, although the decision makers will make discoveries on their own, a facilitator can help them explore productively. At the beginning of an SDILE session, a clear description of the goals and guidance about how to form and test hypotheses would benefit the decision makers' exploration experiences.

This level of conceptual decomposition of facilitator support is critical to the design of an SDILE to support decision making and learning in complex dynamic environments. However, it has been unfortunately unattended and even ignored by the literature of dynamic decision making and interactive learning environments, most of which equate facilitator support with just "sign-in" introduction remarks and/or a half-hearted post-task "thank you" words but fail to explore the underlying components—pre-task, in-task, and post-task level facilitator support.

This reflection on facilitator support from cognitive apprenticeship theory, together with the conceptual exploration of facilitator support, further stimulates the current research: How should facilitator support be designed in order to help decision makers more effectively acquire structural knowledge and/or develop useful heuristic and hence improve task performance? It is worth noting that Klabbers' learning model (Klabbers 2000) is incongruent with Collins' apprenticeship theory: they complement each other and provide stronger support for the design of an effective SDILE.

2.2.2   Gagné's Learning Theory

Gagné's learning theory for instructional design is particularly useful for the design of a learning environment aimed at improving the user's decision making in dynamic tasks (Spector 2000; Qudrat-Ullah 2010). Gagné's learning principles are based on knowledge of how human beings process information. These learning principles refer to a systematic set of actions that has to occur in a learning environment for effective knowledge acquisition and transfer (Gagné et al. 1992). Gagné suggested nine events of instruction for the design of an interactive learning environment such as SDILE that may enhance user's learning and decision making in the following ways:

1. Gain attention.
2. Inform learners of objectives.
3. Stimulate recall of prior learning.

Table 2.2 Incorporation of *Gagné*'s nine instructional events

Instructional event (learning objective)	Implementation in SIADH-ILE
1. Attention and arouse interest and curiosity (reception)	The player at the very first screen is presented with a "challenging task" with the help of "text window" and background pictures of relevant scenes to gain his/her *attention* and arouse interest in the dynamic task
2. Clarity of the object of the ILE (expectancy)	The *objective* is presented in clear terms in SIADH-ILE; "How to use resources effectively for HIV/AID treatment and prevention?" Also, briefing contributes here
3. Recall of prior knowledge (retrieval)	The "pre-playtest" helps stimulates the *recall of prior knowledge*
4. Material presentation (selective perception)	Text and objects are used in SIADH-ILE *for material presentation*
5. Guidance for learning (semantic encoding)	Our SIADH-ILE has two runs. After the first run, the players are led to an explanation interface where they are presented with a generic causal loop diagram together with examples as *guidance for learning*
6. Performance elicitation (responding)	The navigational buttons of the SIADH-ILE allow the users to go back and forth from generic to specific explanations and vice versa: facilitating the *performance elicitation*
7. Provide feedback (reinforcement)	The explanation interface also facilitates the analyses of the players' case, proving the *feedback* before the second run. Also, the pop-up window messages *provide feedback*
8. Assess the performance (retrieval)	The post-playtest is designed to *assess the performance* of the player
9. Enhancing retention and transfer of knowledge (generalization)	In SIADH-ILE, the structure-behavior graphs provide useful insights applicable to other real-world situations. The debriefing session augments the last instructional event: *enhancing retention and transfer of knowledge*

4. Present stimulus.
5. Provide learner guidance.
6. Elicit performance.
7. Provide feedback.
8. Assess performance.
9. Enhance retention and transfer.

Table 2.2 shows how we have implemented these instructional events in the SIADH-ILE.

2.3   Use of Structured Debriefing in the Design of SDILEs

All simulation-based education and learning are experiential in nature. Users take actions or make decisions, task system changes, and feedback is provided to the users to allow them to make their next decisions. The cycle continues until the end of the session with the simulated task environment. Users might have rich learning experiences with the simulation environment. For effective learning to occur, learners[1] have to overcome several challenges:

(i) *Inadequacies of their mental models*

The behavior of dynamic task systems is often counterintuitive (Sterman 2000). After having made several decisions and observing dynamic feedback in the simulated environment, learners face difficulty in knowing what actually did happen: how did their decisions lead to [this] performance. In fact, learners have some preconceived beliefs or mental model about the task at hand, e.g., what types of decision strategies are needed to achieve the desirable outcome (e.g., minimizing the HIV/AIDS-related deaths in our SIADH-ILE's underlying dynamic task). However, when they play out their decision strategies in the simulated task, often the outcome is not what they perceived, even though they thought they understood the interactions in the task system.

(ii) *Structural learning challenge*

In dynamic tasks, an appreciation and understanding of the basic principle of system dynamics (Forrester 1961): that structure drives performance, is critical. Also, learners, after the simulation experience, need to assimilate the learned new knowledge with the existing knowledge they have.

(iii) *Cognitive overload challenge*

In ILEs, the underlying dynamic tasks are cognitively intensive. Learners face the task of processing of overwhelming information to draw lessons and develop useful insights. They need a support mechanism to reduce this cognitive load and build useful insights.

(iv) *Transfer learning challenge*

The key purpose of all simulation-based training is to enable the learners to acquire skills and knowledge to solve real-world problems (e.g., if a person flies a plane or performs a surgery, etc.). They, initially just after the simulation session, have no way of knowing which are important and useful in the real world.

To improve people's decision making in dynamic tasks, learners can't be left alone after accomplishing the dynamic task in the simulated environment. To overcome the abovementioned challenges, therefore, a structured debriefing activity is needed. A debriefing is post-[simulation]-experience feedback-oriented interactive session whereby the decision makers are provided with in-depth facilitation and reflection on their decision making experiences. It aims to improve their decision making skills in cognitively intensive, dynamic tasks (Fanning and Gaba 2007; Lederman 1992; Qudrat-Ullah 2014). According to Lederman, debriefing is the processing of simulation-based learning experience from which the decision makers are to draw the lessons to be learned (Lederman 1992). Consider the case of a trainee surgeon. After intensive simulation-based training and supervised, often under a senior team of surgeons, hands-on experience in the operation theater, a

---

[1] Users and learners are used intermittently throughout this book.

trainee surgeon has to be debriefed. In this postexperience session, her/his performance is reviewed, analyzed, critiqued, gratified, and reflected upon, and any misconceptions are clarified by the expert surgeon (s) (Tannenbaum and Cerasoli 2013). According to Lederman (1992), there are seven common structural elements involved in the debriefing process: (i) debriefer, (ii) participants to debrief, (iii) some kind of experience (simulation session), (iv) the impact of the simulation experience, (v) recollection, (vi) report, and (vii) time. In our process-oriented debriefing session, with the availability of structure-behavior graphs about the various variables of the dynamic task system, the debriefer was equipped well to incorporate these structural elements of the debriefing.

One should note that debriefing and feedback are not the same. Table 2.3 presents the salient characteristics of debriefing and feedback sessions.

To improve the effectiveness of simulation-based education and learning, various efforts have been made to design and implement various debriefing protocols and frameworks. For instance, Thatcher and Robinson (1985) designed a debriefing protocol which focusses on the following aspects:

(i) Impact of the simulation experience,
(ii) Process development
(iii) Clarification of the facts, concepts, and principles
(iv) Emotion involvement
(v) Various views formed by the participants

On the other hand, Lederman's model of debriefing emphasizes that a debriefing session has to be effectively accomplished in three learning enhancing activities, which are as follows:

(i) Analysis of (simulation-based decision making) experience
(ii) Personalization of experience
(iii) Generalization of experience

Consistent with the general objective of a learning session, this protocol helps the learners to not only understand the structure of the task but also develop transferable skills.

Table 2.3 Debriefing versus feedback

Debriefing characteristics	Feedback characteristics
Debriefing sessions are well-engaged and facilitate active learning	Feedback often provides passive learning
Debriefing sessions focus on not only what happened but also why it happened	In feedback sessions, corrective feedback often forms the basis of discussion
Debriefing sessions are learning improvement oriented	Feedback sessions are assessment oriented
Debriefing sessions are focused on reflections on individual performance, peers performance, and expert performance	Feedback sessions rarely go beyond self-reflection or appraisal

Toward the design and implementation of debriefing to enhance users' learning and decision making in dynamic tasks, system dynamics researchers have also suggested several protocols. For instance, Pavlov et al. (2015) proposed a pedagogical framework called Game-Based Structural Debriefing (GBSD) which supplements a video game, Food Fight, with structural debriefing, which can support students' learning about the internal causal structure of a complex system portrayed in a simulation or a video game (Pavlov et al. 2015). In this protocol, students are facilitated to relate game decision making to task system concepts (e.g., causal relationships, feedback, accumulation, and delay). Using GBSD framework, Kim and Pavlov (2017) implemented a game-based curriculum. It addresses the question of how system dynamics practices could be used by teachers as a design tool (Kim and Pavlov 2017). They found mixed results: students were able to identify the causal relationships between task variables but found difficulty in understanding the dynamic behavior of feedback loops.

With regard to the implementation of debriefing in our ILE, SIADH-ILE, we have benefited from these earlier developments in the debriefing design area. Our SIADH-ILE, specifically, draws on Lederman's model, where a debriefing session has effectively accomplished three learning enhancing activities:

(i)   Analysis of (simulation-based decision making) experience
(ii) Personalization of experience
(iii) Generalization of experience

With the use of SIADH-ILE-based outputs of users and the structure-behavior graphs of the help system of SIADH-ILE, the facilitator conducts a structured briefing following Lederman's model.

Debriefing can be delivered in different forms, shapes, and methods. For instance, oral discussions, written notes, and debriefing games are the most common forms of debriefing (Lederman 1992; Peters and Vissers 2004). In an oral discussion, learners and debriefs engage in a question and answer session designed to guide learners through a reflective process about their learning. In written notes, a passive form of debriefing, the learners are provided with handouts that present "expert solution" to the task they had in the ILE and examples of potential applications of their learning. Use of causal loop diagrams is common in SDILE-based debriefing sessions (Qudrat-Ullah 2010). Debriefing games are interactive strategies, played through a computer or board games where the learners are encouraged to reflect on earlier events (Thiagarajan 1992). In interactive oral discussion type of debriefing, the interactive sessions can be organized in two ways: (i) where participants are presented with a sort of "expert solution" to the task in the ILE and are asked to recall, reflect, and compare their "own" solutions (Lederman 1992; Peters and Vissers 2004) and (ii) where participants are led through a process that illustrates the underlying structure of the task systems and how it relates to the behavior of the task system (Cox 1992; Crook-All et al. 1987). We term former as "outcome-oriented debriefing" and later as "process-oriented debriefing." This distinction is important because in well-structured tasks, outcome-oriented debriefing alone may be sufficient to improve task performance. However, in a dynamic task which embodies time delays, feedback loops, and uncertainties (e.g., firefighting, flying a plane, and performing surgeries in emergency wards), outcome-oriented debriefing is not enough (Blazer et al. 1989; Lakeh and Ghaffarzadegan 2015). Instead, a process-oriented debriefing, where causal relationships between task variables are explicitly elaborated, is needed. One should help the learners to overcome the misperceptions about the task. Research shows that learning and task performance improve when participants in management exercises understand the structure of the system they control (Isaacs and Senge 1994; Sterman 2000; Qudrat-Ullah 2007). Thus, here in the case of process-oriented debriefing, it is not a question of more information but the specific nature of the information: structure-behavior graphs elucidating causal information versus expert solution-based information. In an outcome-oriented debriefing session, where expert solutions become the base of learning activities, subjects can resort to a relatively easy, random decision strategy. However, when learners are trained in a process-oriented debriefing session, they are expected to follow systematic and causal learning-oriented decision strategies allow the users to develop heuristics knowledge and structural knowledge. Following systematic decision strategies, users can become efficient in decision making. Instead of indulging in guessing like those applying random strategies, they can make use of learned causal relationships and become efficient decision makers. Also, debriefing plays a fundamental role in helping the participants connect the knowledge and skills developed in a simulation session to the corresponding real-life situation—transfer learning (Dreifuerst 2009; Qudrat-Ullah 2014; Lane and Tang 2000). Therefore, we assert that learners in a debriefing-based SDILE will have the opportunity to develop such transfer learning skills.

2.3.1 Debriefing-Based SDILE and Task Performance

Most of the research on dynamic decision making (DDM) and learning in SDILEs has focused on improving managerial performance (Bakken 1993; Langley and Morecroft 1995; Moxnes 2004). The key finding of DDM literature is that people perform poorly in dynamic tasks due to the misconception of the dynamic task, also known as MOF hypothesis: they fail to understand the effects of time delays, delayed feedback, and nonlinear relationships among the variables of the dynamic task (Brehmer 1990; Sterman 1994). However, to perform better in dynamic tasks, decision makers need to develop an adequate model of the task system (Sterman 2000). SDILEs provide a practice field for decision makers to develop a better understanding of the dynamic tasks. However, practice with SDILEs alone is not sufficient in developing a better understanding of the dynamic task. In a SDILE, after an experience with a simulated task environment, subjects have rich learning experiences. These subjects face the cognitively intensive task of processing of overwhelming information to draw lessons and develop useful insights. They need instructional support that reduces this cognitive load (Sweller 1988). Debriefing sessions when subjects are provided with guidance and support in (i) relating the learned knowledge to real-world situations and (ii) clarifying any misconceptions about the task system (Briggs 1990; Sterman and Booth Sweeney 2007) are expected to provide this support.

In dynamic tasks, outcome-oriented debriefing does not provide enough information to the participants to enable them to form a suitable model of the dynamic task (Blazer et al. 1989; Sterman 1989; Lakeh and Ghaffarzadegan 2015). Individuals need to understand both the delays and the feedback structures underlying the task. Process-oriented debriefing, however, has the potential to impart this crucial knowledge: the debriefer identifies the feedback structures and their relation to the outcomes, delays are examined, and uncertainties are discussed. Several structure-behavior graphs (e.g., see Fig. 2.2) are presented, and the relationships among the key variables of the dynamic task are made explicit. Provision of causal loop diagrams of the dynamic task improves subjects' causal understanding of the task (Rouwette et al. 2004). Consequently, increased causal understanding of a dynamic task enables the decision makers to perform better (Plate 2010; Qudrat-Ullah 2014).

Overall, process-oriented debriefing may help the decision makers in at least two ways: (i) expanding the limits of their cognitive capacity (e.g., by already having an understanding of the task structures, cognitive resources are freed-up for other uses) and (ii) reducing their misconceptions about the dynamic task (i.e., they have a better understating about structure-behavior relationships). With increased cognitive capacity, a decision maker is able to process more information, evaluate more alternatives, and perform better (Hogarth and Makridakis 1981; Qudrat-Ullah 2014). Therefore, we assert that subjects with debriefing should perform better in dynamic tasks.

2.3.2   Debriefing and Decision Strategy

When it comes to decision making, be it in a static environment or in a dynamic environment, subjects adopt to specific decision strategies. Researchers have found that the decision making environments shape the decision strategies of the decision makers (Hogarth and Makridakis 1981; Kleinmuntz 1985; Größler et al. 2016). Faced with the varying complexities of dynamic tasks, the decision maker can adopt to several decision strategies. Dynamic tasks by their nature are uncertain and unpredictable (Sterman 2000; Kwakkel and Pruyt 2013). Research has shown that when tasks have uncertainties, people often resort to anchoring and adjustment strategy or arbitrary random strategies to do the task (Tversky and Kahneman 1974; Rouwette et al. 2004). In an anchoring and adjustment strategy, decisions are based on anchoring on the previous decision and adjusting by a certain percentage. Although such a strategy has some utility, e.g., in simple choice and judgmental tasks, success in dynamic tasks requires better strategies (Größler et al. 2016). These strategies, to the minimum, should allow the decision makers to trace and recognize the patterns of the dynamic tasks.

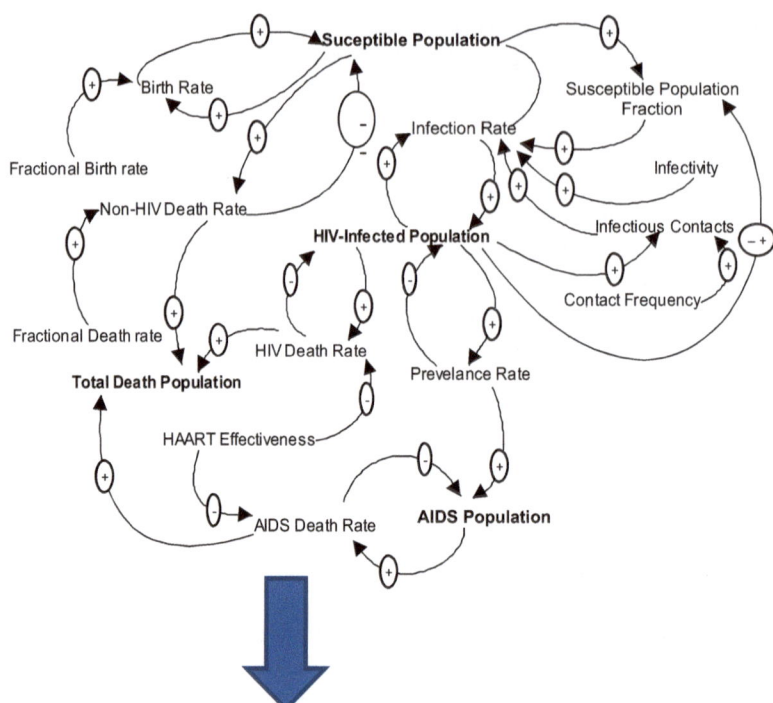

Effect of Contact Frequency Improvements on AIDS and HIV-Infected Populations

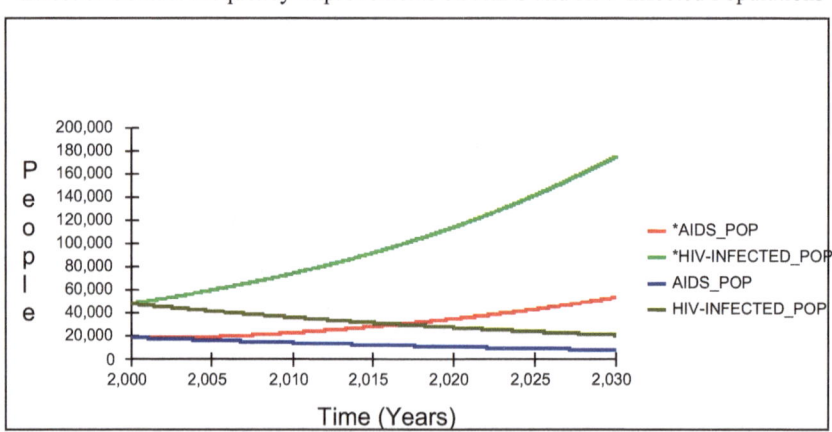

Fig. 2.2 An example of structure-behavior graph. Legend names with "*" represent the baseline scenario. Here, on the top side, we see feedback loop structures that are responsible for creating the behavior of AIDS and HIV-infected population as shown in the graph above

Sengupta and Abdel-Hamid (1993) in their experimental research found that subjects provided with cognitive feedback improved their decision making capabilities by enhancing their comprehension of the task structure, employed consistent decision strategies, and performed better than those provided with outcome feedback alone. With outcome feedback alone, inferring relationships between the variables of a dynamic task is a challenge. Process-oriented debriefing, with the potential to aid the decision makers to develop a suitable model of the dynamic task (Conant and Ashby 1970; Fischer and Gonzalez 2016), should induce decision makers to adopt exploratory decision strategies. These exploratory strategies will facilitate the pattern recognition abilities of the decision makers, and hence they are expected to perform better. In process-oriented debriefing sessions, the structure-behavior relationship of the task system with the help of structural diagrams such as shown in Fig. 2.2 is explicitly elaborated. Therefore, with an increased knowledge about the structure-behavior relationship of the task system at hand, decision makers are more likely to adopt a systematic variable strategy (SVS) rather than a random strategy. In an SVS, the optimal values of decision variables will vary over the various decision periods accounting for the changes in central variables of the task system. Therefore, we put forward an assertion that subjects with process-oriented debriefing will adapt to the relatively more systematic variable strategy.

2.3.3   Debriefing and Decision Time

In organizational decision making, both the decision accuracy and decision efficiency are needed. SDILEs, like any decision support tool, affect the decision making process by making it more or less efficient. Process-oriented debriefing fosters decision makers' decision efficiency by at least two ways: (i) allowing them to reflect on their performance and experiences with the ILE (Briggs 1990; Sterman 2000; Qudrat-Ullah 2007) and (ii) making them aware of their entrenched beliefs and facilitating their eventual reinterpretation (Faria 1998; Kottermann et al. 1995; Qudrat-Ullah 2014). Therefore, process-oriented debriefing can reduce decision making time by helping users make judgmental inputs and avoid spending extra time searching for, e.g., which variables of a task system give rise to specific system behavior. Drawing upon the cognitive load theory that argues that instructional support can be designed to reduce the cognitive load of the learners (Sweller 1988), availability of task-related information (e.g., structure-behavior graphs as shown in Fig. 2.2) is expected to reduce the cognitive load of the decision maker. Moreover, in probabilistic tasks (Glöckner and Betsch 2008; Söllner et al. 2013) and time-critical tasks (Cohen 2008), accessibility of task-related information leads to automatic decision making. There is no reason at priori not to believe that the availability of process-oriented debriefing will make the decision makers more efficient decision makers. On the other hand, users of SDILEs with outcome-oriented debriefing must turn solely to their cognitive resources to infer the relationships between key task variables (Sengupta and Abdel-Hamid 1993), leading to more time to digest the information and to make decisions. Therefore, we assert that the users of an SDILE with process-oriented debriefing will become relatively more efficient decision makers.

2.3.4 Debriefing and Decision Structural Knowledge

Structural knowledge pertains to knowledge about principles, concepts, and facts about the underlying model of a decision task (Berry 1991). In most of the organizational tasks, decision makers have to develop the structural knowledge of the task system to perform better. In the context of a dynamic task, Berry (1991) found that participants improved both their structural knowledge and their task performance when tasks were changed with relatively more transparent relations among the task variables. When decision makers are provided with structural information about the underlying dynamic task, their task knowledge improves (Sterman 2000; Qudrat-Ullah 2007). Unlike the outcome-oriented debriefing where the focus is on sharing of expert solutions to the participants, a process-oriented debriefing provides an opportunity for a facilitator to elaborate on the relations between key variables of the dynamic task (e.g., with the help various structure-behavior graphs). As a result, decision makers may develop a better understanding of the facts and principles (e.g., which variables are related to the goal-attaining behavior of the task system) about the underlying task. The subjects with outcome-oriented debriefing might have just mimicked the shown expert solution. Consequently, users of SDILEs with process-oriented debriefing should acquire better structural knowledge. Therefore, we propose that users with SDILEs with process-oriented debriefing will develop relatively more structural knowledge as compared with those with outcome-oriented debriefing or with no debriefing.

2.3.5 Debriefing and Decision Heuristics Knowledge

Heuristics knowledge concerns how decision makers actually control or manage a task (e.g., flying a plane, putting out a fire, performing an ER operation, etc.) (Qudrat-Ullah 2014). Consider the example of bicycle riding. No matter how good a manual is provided or how good instructions are given by even the best expert, the learner will not be able to ride the bike. Instead, she/he should just do it: riding, falling, and riding. Then, the learners will develop some useful heuristics to ride the bike. Dynamic decision makers may well know the strategies to achieve better task performance (heuristics knowledge), even though they cannot show improvement in declarative knowledge or structural knowledge (Berry 1991). In an outcome-oriented debriefing session, the availability of and discussions on the expert solution to the dynamic task might help the decision makers to develop heuristics knowledge. However, a process-oriented debriefing session where participants are provided with information on causal relationships between task variables may enhance their capability for dynamic decision making. Thus, users of SDILEs with process-oriented debriefing are expected to acquire more heuristics knowledge.

2.4   Measurement of Decision Making and Learning with SDILEs

Prior studies on DDM have often used game score or task performance as the measure of learning outcome. We have considered two kinds of decision outcomes: decision making and learning. In this study, consistent with prior studies on dynamic tasks, subject's decision making is conceptualized as user's ability to do the task (Abdel-Hamid et al. 1999; Sterman 1994; Größler et al. 2016). Maximizing the profit of a firm or the minimization of its cost or completion of a task in minimum time is commonly used as the example of task performance measures. Thus, task performance, the user's decision strategies and their efforts (i.e., decision efficiency) are the measures for subjects' decision making.

However, the other decision outcome is learning, bear on multiple perspectives. For instance, we may consider learning either as a progression toward expertise or as becoming part of a community of practitioners (Spector 2000). Following the modern objective-oriented "constructivist" approach to learning, we adapt to Sternberg's view of learning (Sternberg 1995). In Sternberg's view, people are not really experts or non-experts, but rather are experts in varying degree—prototypes. A prototype view of expertise implies a broader view of learning and can accommodate the diversity of skills and knowledge acquired by decision makers through participation in SDILE-based decision making sessions. Following the theoretical framework of Sternberg and Horvath (1995), decision makers in SDILEs can develop, for example, varying levels of task knowledge and insights regarding decision making in dynamic tasks. To better capture a range of such expertise development, therefore we have applied a five-dimensional measurement model: task performance, decision strategy, decision time, structural knowledge, and heuristics knowledge. Therefore, improved performance in dynamic tasks could be evident from an improvement in any of these dimensions of expertise (Yi and Davis, 2001). A list of metrics into decision processes and decision outcomes (Qudrat-Ullah 2007, 2014) to evaluate the effectiveness of debriefing-based SDILE on subject's decision making and learning in dynamic tasks is shown in Table 2.4.

2.5   A Conceptual Model of Decision Making and Learning with SDILEs

Education and training with SDILEs are expected to improve people's decision making and learning in dynamic tasks. In the context of our proposed multidimensional evaluation model which takes into account varying level of expertise development (Sternberg 1995), we specifically propose the following set of testable assertions:

Table 2.4 Metrics to assess the effectiveness of debriefing-based SDILEs

Concepts	Decision processes (What is the nature of subjects' decision processes in terms of consistency and efficiency?)	Decision outcomes	
		Decision making (What happened? How did a subject do the task?)	Learning (To what extent a subject acquired task knowledge?)
Metrics	Decision strategy (consistent or random strategy)	Task performance (% deviation from the benchmark)	Structural knowledge (# of correct answers)
	Decision time (Average amount of time it takes to make a decision)		Heuristics knowledge (# of correct answers)

Source: Adapted from (Hassan, SDR)

H_1: Debriefing will help the learner to improve their performance in the dynamic task.

H1a: Process-oriented debriefing will be more effective than outcome-oriented debriefing in improving learner's performance in the dynamic task.

H_2: Debriefing will help the learners develop structural knowledge.

H2a: Process-oriented debriefing will be more effective than outcome-oriented debriefing in improving learner's structural knowledge.

H_3: Debriefing will help the learners develop heuristics knowledge.

H3a: Process-oriented debriefing will be more effective than outcome-oriented debriefing in improving learner's heuristics knowledge.

H_4: Debriefing will help the learners become efficient decision makers.

H4a: Process-oriented debriefing will be more effective than outcome-oriented debriefing in improving learner's decision making efficiency.

H_5: Debriefing will help the learners develop systematic decision strategies.

H5a: Process-oriented debriefing will be more effective than outcome-oriented debriefing in developing learner's systematic decision strategies.

Figure 2.3 represents the overall hypothesized relationships linking the decision maker, decision environment (i.e., SDILE), and performance measurements (i.e., decision making and learning).

2.6 Summary of the Model for the Effects of Debriefing on People's Decision Making

To improve the efficacy of decision aids such as SDILEs, the design basis of SDILEs matters. Drawing upon Gagné's nine instructional events, Collin's apprenticeship theory, HCI design principles, and Lederman's debriefing model, we have designed and developed an SDILE, SIADH-ILE. To evaluate the effectiveness of SIADH-ILE or any other SDILE and better capture the learning and decision making of the users of SDILEs, we have utilized Sternberg's view of "prototypic expertise" to develop a five-dimensional model. Subject's decision making will be measured

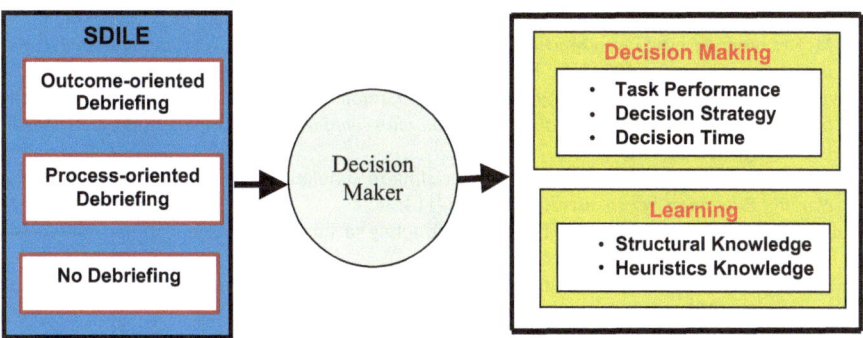

Fig. 2.3 The effects of debriefing on decision making and learning

through three dependent variables: (i) task performance, decision time, and (ii) decision strategy. Learner's learning is captured through two independent variables: (i) structural knowledge and (ii) heuristics knowledge.

Overall, a new conceptual model for the effects of debriefing (i.e., process-oriented and outcome-oriented debriefing) on people's decision making and learning in dynamic tasks is proposed here that explicitly integrates the decision maker, the decision making environment, and decision making performance measurements. In the next chapter, we will provide the details about the laboratory experiments that are conducted to empirically test this model.

References

Abdel-Hamid, T., Sengupta, K., & Swett, C. (1999). The impact of goals on software project management: An experimental investigation. *MIS Quarterly, 23*(4), 1–19.

Bakken, B. E. (1993). Learning and transfer of understanding in dynamic decision environments. Unpublished doctoral dissertation, MIT, Boston.

Berry, D. C. (1991). The role of action in implicit learning. *Quarterly Journal of Experimental Psychology, 43A*, 881–906.

Blazer, W. K., Doherty, M. E., & O'Connor, R. (1989). Effects of cognitive feedback on performance. *Psychological Bulletin, 106*(3), 410–433.

Brehmer, B. (1990). Strategies in real-time dynamic decision making. In R. M. Hogarth (Ed.), *Insights in decision making* (pp. 262–279). Chicago: University of Chicago Press.

Briggs, P. (1990). Do they know what they are doing? An evaluation of word-processor user's implicit and explicit task-relevant knowledge, and its role in self-directed learning. *International Journal of Man-Machine Studies, 32*, 385–298.

Cohen, I. (2008). Improving time-critical decision making in life-threatening situations: Observations and insights. *Decision Analysis, 5*(2), 100–110.

Crookall, D., Oxford, R., & Saunders, D. (1987). Towards a reconceptualization of simulation: From representation to reality. *Simulation & Gaming, 17*(4), 141–171.

Collins, A. (1991). Cognitive apprenticeship and instructional technology. In L. Idol & B. F. Jones (Eds.), *Educational values and cognitive instruction: Implications for reform* (pp. 121–138). Hillsdale, NJ: Lawrence Erlbaum Associates, Inc..

Conant, R., & Ashby, W. (1970). Every good regulator of a system must be a model of the system. *International Journal of System Science, 1*, 89–97.

Cox, R. J. (1992). Exploratory learning from computer-based systems. In S. Dijkstra, H. P. M. Krammer, & J. J. G. van Merrienboer (Eds.), *Instructional models in computer-based learning environments* (pp. 405–419). Berlin, Heidelberg: Springer-Verlag.

Davidsen, P. I., & Spector, J. M. (1997). Cognitive complexity in system dynamics based learning environments. *International system dynamics conference*, Istanbul, Turkey: Bogazici University Printing Office, 757–760.

Dreifuerst, K. T. (2009). The essentials of debriefing in simulation learning: A concept analysis. *Nursing Education Perspectives, 30*(2), 109–114.

Fanning, M., & Gaba, M. (2007). The role of debriefing in simulation-based learning. *Simulation in Healthcare, 2*(2), 115–125.

Faria, A. J. (1998). Business simulation games: Current usage levels—An update. *Simulation and Gaming, 29*, 295–308.

Fischer, H., & Gonzalez, C. (2016). Making sense of dynamic systems: How our understanding of stocks and flows depends on a global perspective. *Cognitive Science, 40*(2), 496–512.

Forrester, J. W. (1961). *Industrial dynamics*. Cambridge, MA: Productivity Press.

Gagné, R. M. (1985). *The conditions of learning and theory of instruction*. New York: Holt, Rinehart, and Winston.

Gagné, R. M., Briggs, L. J., & Wager, W. W. (1992). *Principles of instructional design* (4th ed.). Forth Worth, TX: Harcourt Brace Jovanovich College Publishers.

Glöckner, A., & Betsch, T. (2008). Multiple-reason decision making based on automatic processing. *Journal of Experimental Psychology: Learning, Memory, and Cognition, 34*, 1055–1075.

Größler, A., Rouwette, E., & Vennix, J. (2016). Non-conscious vs. deliberate dynamic decision-making—A pilot experiment. *Systems, 4*(13), 1–13. https://doi.org/10.3390/systems4010013.

Hogarth, R. M., & Makridakis, M. (1981). Beyond discrete biases: Functional and dysfunctional aspects of judgmental heuristics. *Psychological Bulletin, 9*(2), 197–217.

Howie, E., Sy, S., Ford, L., & Vicente, K. J. (2000). Human-computer interface design can reduce misperceptions of feedback. *System Dynamics Review, 16*(3), 151–171.

Isaacs, W., & Senge, P. (1994). Overcoming limits to learning in computer-based learning environments. In J. Morecroft & J. Sterman (Eds.), *Modeling for learning organizations* (pp. 267–287). Portland, OR: Productivity Press.

Kim, J., & Pavlov, O. (2017). Game-based structural debriefing: A design tool for systems thinking curriculum. *SSRN Electron J.* https://doi.org/10.2139/ssrn.3218674.

Klabbers, G. (2000). Gaming & simulation: Principles of a science of design. *Simulation and Gaming, 34*(4), 569–591.

Kleinmuntz, D. (1985). Cognitive heuristics and feedback in a dynamic decision environment. *Management Science, 31*, 680–701.

Kottermann, E., Davis, D., & Remus, E. (1995). Computer-assisted decision making: Performance, beliefs, and illusion of control. *Organizational Behavior and Human Decision Processes, 57*, 26–37.

Kwakkel, J. H., & Pruyt, E. (2013). Explanatory modeling and analysis and approach for model-based foresight under deep uncertainty. *Technological Forecasting and Social Change, 80*(3), 419–431.

Lakeh, B., & Ghaffarzadegan, N. (2015). Does analytical thinking improve understanding of accumulation? *System Dynamics Review, 31*(1–2), 46–65.

Lane, M., & Tang, Z. (2000). Effectiveness of simulation training on transfer of statistical concepts. *Journal of Educational Computing Research, 22*(4), 383–396.

Langley, P. A., & Morecroft, J. D. W. (1995). Learning from microworld environments: A summary of the research issues. In G. P. Richardson & J. D. Sterman (Eds.), *System dynamics '96* (pp. 213–231). Cambridge, MA: System Dynamics Society.

Lederman, L. C. (1992). Debriefing: Towards a systematic assessment of theory and practice. *Simulation and Gaming, 23*(2), 145–160.

Mayer, W., Dale, K., Fraccastoro, K., & Moss, G. (2011). Improving transfer of learning: Relationship to methods of using business simulation. *Simulation and Gaming, 42*(1), 64–84.

Moxnes, E. (2004). Misperceptions of basic dynamics: The case of renewable resource management. *System Dynamics Review, 20,* 139–162.

NTSA. (2011). President's notes. *Training Industry News, 23*(4), 2–2.

Pavlov, O., Saeed, K., & Robinson, L. (2015). Improving instructional simulation with structural debriefing. *Simulation and Gaming, 46*(3–4), 383–403.

Peters, V. A. M., & Vissers, G. A. N. (2004). A simple classification model for debriefing simulation games. *Simulation & Gaming, 35*(1), 70–84.

Plate, R. (2010). Assessing individuals' understanding of nonlinear causal structures in complex systems. *System Dynamics Review, 28*(1), 19–33.

Qudrat-Ullah, H. (2007). Debriefing can reduce misperceptions of feedback hypothesis: An empirical study. *Simulations and Gaming, 38*(3), 382–397.

Qudrat-Ullah, H. (2010). Perceptions of the effectiveness of system dynamics-based interactive learning environments: An empirical study. *Computers and Education, 55,* 1277–1286.

Qudrat-Ullah, H. (2014). Yes we can: Improving performance in dynamic tasks. *Decision Support Systems, 61,* 23–33.

Rouwette, A., Großler, A., & Vennix, M. (2004). Exploring influencing factors on rationality: A literature review of dynamic decision-making studies in system dynamics. *Systems Research and Behavioral Science, 21,* 351–370.

Sengupta, K., & Abdel-Hamid, T. (1993). Alternative concepts of feedback in dynamic decision environments: An experimental investigation. *Management Science, 39,* 411–428.

Söllner, A., Brödery, A., & Hilbig, E. (2013). Deliberation versus automaticity in decision making: Which presentation format features facilitate automatic decision making? *Judgment and Decision making, 8*(3), 278–298.

Spector, J. M. (2000). System dynamics and interactive learning environments: Lessons learned and implications for the future. *Simulation and Gaming, 31*(4), 528–535.

Sterman, J. D. (1989). Modeling managerial behavior: Misperceptions of feedback in a dynamic decision making experiment. *Management Science, 35,* 321–339.

Sterman, J. D. (1994). Learning in and about complex systems. *System Dynamics Review, 10,* 291–330.

Sterman, J. D. (2000). *Business dynamics: Systems thinking and modeling for a complex world.* New York: McGraw-Hill.

Sterman, J. D., & Booth Sweeney, L. (2007). Understanding public complacency about climate change: Adults' mental models of climate change violate conservation of matter. *Climatic Change, 80*(3–4), 213–238.

Sternberg, R. J. (1995). Expertise in complex problem solving: A comparison of alternative conceptions. In P. Frensch & J. Funke (Eds.), *Complex problem solving: The European perspective* (pp. 3–25). Hillsdale, NJ: Lawrence Erlbaum Associates Publishers.

Sternberg, R. J., & Horvath, J. A. (1995). A prototype view of expert teaching. *Educational Researcher, 24*(6), 9–17.

Sweller, J. (1988). Cognitive load during problem solving: Effects on learning. *Cognitive Science, 12*(2), 257–285.

Tamara, E. F., Alec, M. B., & Judith, D. S. (2013). The use of technology by nonformal environmental educators. *The Journal of Environmental Education, 44*(1), 16–37.

Tannenbaum, S. I., & Cerasoli, C. P. (2013). Do team and individual debrief enhance performance? A meta-analysis. *Human Factors, 55*(1), 231–245.

Thatcher, C., & Robinson, J. (1985). *An introduction to games and simulations in education. Simulations.* Hants: Solent.

Thiagarajan, S. (1992). Using games for debriefing. *Simulation & Gaming, 23*(2), 161–173.

Tversky, A., & Kahneman, D. (1974). Judgment under uncertainty: Heuristics and biases. *Science, 185*(4157), 1124–1131.

Yi, S., & Davis, F. (2001). Improving computer training effectiveness for decision technologies: Behavior modeling and retention enhancement. *Decision Sciences, 32*(3), 521–544.

Chapter 3
The Experimental Approach

Abstract To empirically test our proposed assertions about the efficacy of debriefing in SDILE-based education and training in dynamic tasks, we adopted the experimental approach. In this chapter, the experimental approach is explained through various dimensions. It has a step-by-step procedure to meet the needs of debriefer and the participants. First, the research design is elaborated. Then, the dynamic task, SIADH-ILE, its causal structure, mathematics model, and interface design including its decision panel and help systems are explained. To capture the knowledge development of the learners, examples of both structural and heuristic questions are provided. Finally, the protocol of the experiment is also described here.

Keywords Effort · AIDS population · Contact frequency · Decision strategy · Decision time · Dynamic task · Experimental approach · HAART · Hawthorne effects · Learning objective · HIV/AIDS epidemic · Scripted discussion · SIADH-ILE · Task performance metric · Transfer learning

3.1 Introduction

We have advanced several assertions in the previous chapter regarding the efficacy of SDILEs in improving people's decision making and learning in dynamic tasks. How are we going to test them empirically? We have adopted the laboratory experimental approach (Moxnes 2004) to allow us to (i) control various factors (e.g., participants' education and background experience) and (ii) perform replicative studies with a single step or factor change in each study. Laboratory experiments are an authentic and scientific method of inquiry. In psychology, medicine, and education fields, use of laboratory experiments is a norm (Moxnes 2004). We are using a pretest—intervention—posttest research design to test our hypotheses.

We developed a task: the management of the HIV/AIDS situation in Canada. This task is dynamic because (i) decision makers have to make a series of interdependent decisions, (ii) several variables have time lags between them (e.g., between prevalence rate and infectivity rate), (iii) it has several nonlinear relationships among variables of the task system, and (iv) it has several feedback loops (Fig. 3.2). The underlying simulation model of the developed SDILE, SIADH-ILE, is a system dynamics model which accounts for all the features of this dynamic task, mentioned above. This simulation model was validated through several structural and behavioral tests (Qudrat-Ullah and BaekSeo 2010; Qudrat-Ullah 2010). Unless a simulation is validated by some acceptable standards, its use to improve one's decision making would be questionable at best.

Our subjects in the experiments are adult learners from the MBA programs of various universities in Toronto, Canada. They were evenly distributed into two experimental groups and one control group. The availability of debriefing (process-oriented or outcome-oriented) was the independent variable, while task performance, decision strategy, decision time, structural knowledge, and heuristic knowledge were the dependent variables of this study. Standard protocols were followed. The participants were provided with monetary incentives. Subjects first practiced with the four scenarios at their disposal and then designed their best policy scenario. Their performance was measured through the count of HIV/AIDS-related deaths and cost (i.e., minimum deaths with the least cost prevention and treatment care is the best performance) throughout 30 years. While data on subjects' task performance, decision strategies, and the decision were automatically recorded by the SIADH-ILE program, the structural and heuristic knowledge was captured through causal-oriented questionnaires. Next, we provide details about the aforementioned aspects of our laboratory experimental approach.

3.2   Research Design

To test the proposed hypotheses, we conducted a laboratory experiment where subjects managed a dynamic task in a controlled simulated environment: SIADH-ILE (Qudrat-Ullah 2007). In this controlled decision making environment, subjects are offered with a "managerial practice field" where they can make a decision and learn in a non-threatening environment. In fact, decision making in these simulation labs is an authentic, enactive mastery experience, whereby the subjects control the operations of their healthcare firms and learn from the outcomes of their decisions (Moxnes 2004). We have chosen an experimental approach to control the hypothesized debriefing that would have influenced the decision makers' decision processes and outcomes.

Figure 3.1 presents our research model. We designed a single factor, completely randomized design involving one control group and two experimental groups, as shown in Table 3.1. Each participant in the experimental group used SIADH-ILE as an SDILE, with either process-oriented or outcome-oriented debriefing. The debriefing was delivered in a scripted discussion between the debriefer and the participants after the participants completed the first formal trial of the task. Each participant

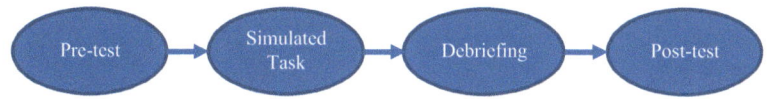

Fig. 3.1 Research model

Table 3.1 Experimental design

	Pretest	Treatment	Posttest
Users with SDILE ($N = 32$)	Yes	Process-oriented debriefing	Yes
Users with SDILE ($N = 32$)	Yes	Outcome-oriented debriefing	Yes
Users with SDILE ($N = 32$)	Yes	No debriefing	Yes

N: Sample size

was asked to agree to a 2.5-hour session. We paid each subject US $50. We assessed the subject's background education, structural knowledge, heuristic knowledge, and demographics through a pretest questionnaire. The computer program embedded in SIADH-ILE allowed the automatic capture of the user's decisions data on decision strategy, decision time, and task performance. A posttest questionnaire and causal loop diagram-based assessment (Spector 2000; Qudrat-Ullah 2014) helped us evaluate subjects' structural and heuristic knowledge improvement if any.

3.3   The Dynamic Task

The HIV/AIDS epidemic in Canada has changed from the early epidemic, which affected primarily men who had sex with men (MSM), to the current epidemic, which increasingly affects other groups such as IDU users, aboriginal people, and women (Atun and Sittampalam 2006). Overall, the HIV/AIDS epidemic among adults in Canada is increasing. Therefore, understanding the dynamics of adults HIV/AIDS situation in Canada has become increasingly important. The need to understand the long-term dynamics of prevention and treatment policy decisions becomes even more pronounced.

Modeling and simulation-based training play a fundamental role in the improvement of healthcare management and delivery across nations (Tebbens and Thompson 2009; Homer and Hirsch 2006; Qudrat-Ullah and Tsasis 2017; Qudrat-Ullah et al. 1997). SIADH, a dynamic model, based on system dynamics which is calibrated to the data of adult HIV/AIDS population in Canada, is developed and applied to better understand the dynamics of the adult HIV/AIDS situation in Canada.

Feedback loops are the fundamental structures of all dynamic tasks. Figure 3.2[1] presents the causal interactions, through various feedback loops, between various

---

[1]An earlier version of SIADH-ILE model description was presented at the 37th International Conference of the System Dynamics Society, Albuquerque, New Mexico, USA ◊ July 21–25, 2019.

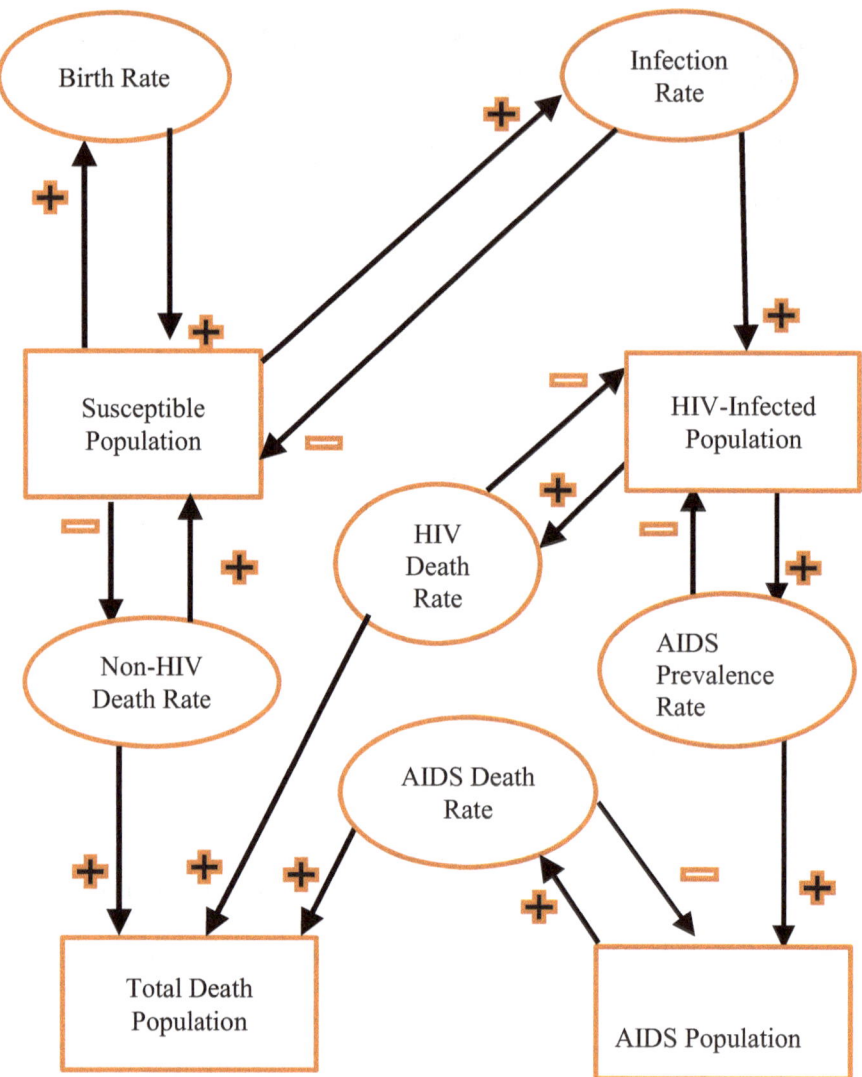

Fig. 3.2 Causal interactions in SIADH-ILE model. (Source: Qudrat-Ullah and Tsasis 2017)

variables of our dynamic task: management of the adult HIV/AIDS prevention and treatment program in Canada.

In the simulation model, SIADH, the susceptible population is dependent on the birth and death rates. The susceptible population comprises of both the HIV-infected and HIV-uninfected adults. The uninfected adults will sooner or later die from other reasons. The HIV-infected individuals are affected through several mechanisms including homosexual contacts, heterosexual contacts, the use of injecting drug users (IDU), and blood transfusion. A higher infection rate leads to a higher infected population and vice versa. For heterosexual adults, preventive measures such as education and awareness can affect the infection rate. Once adults are HIV-infected, without any treatment, they can develop full-blown AIDS within 10 years. Those in treatment such as on highly active antiretroviral therapy (HAART) will have a relatively higher post-HIV survival time. On the other hand, some HIV-infected adults will die without being in either of these two situations and will add to the total death stock. For a decision maker to make better decisions in such a task such as SIADH-ILE, the understanding of the structures of this model is essential. For instance, the availability of better medicine, HAART, takes time.

Specifically, the US Federal Drug Administration (FDA) approval process has a significant time lag. So, the users of SIADH-ILE have to account for such delays else they will be making poor decisions. Likewise, the relationship between contact frequency and infectivity rate is hardly a linear relationship. Instead, it behaves more like a quadratic function: initially few people are infected, so contact frequency is relatively small. However, when more people get infected, the frequency picks up rapidly. Thus, recognizing this kind of non-linear relationship between variables is critical for successful decision making about this dynamic task.

Overall, in the simulation model, SIADH, there are four accumulations (stocks): the susceptible population, the HIV-infected population, the AIDS population, and the total death population. These stocks are responsible for key dynamics of resources for the prevention and treatment program, over time. The mathematical equations of these four stocks are given below:

$$\frac{\partial PS(t)}{\partial(t)} = R_b PS(t) - R_i PS(t) - R_{nd} PS(t) \tag{3.1}$$

$$\frac{\partial PI(t)}{\partial(t)} = R_i PI(t) - R_p PI(t) - R_h PI(t) \tag{3.2}$$

$$\frac{\partial PA(t)}{\partial(t)} = R_p PI(t) - R_a PA(t) \tag{3.3}$$

$$\frac{\partial PD(t)}{\partial(t)} = R_{ni} PS(t) + R_h PI(t) - R_a PA(t) \tag{3.4}$$

where PS (t) = susceptible population, PI (t) = infected population, PD (t) = death stock, R_b = birth rate, R_i = infection rate, R_{nd} = non-HIV death rate, R_p = prevalence rate, R_h = aids death rate, and R_a = infection death rate.

Figure 3.2 shows the control panel and some of the graphs the users can use to interact with the simulation model. The users enter their decisions through each scenario. Once a scenario is specified, users can then execute their decisions by clicking the button "PLAY." The users also have the opportunity to view the results

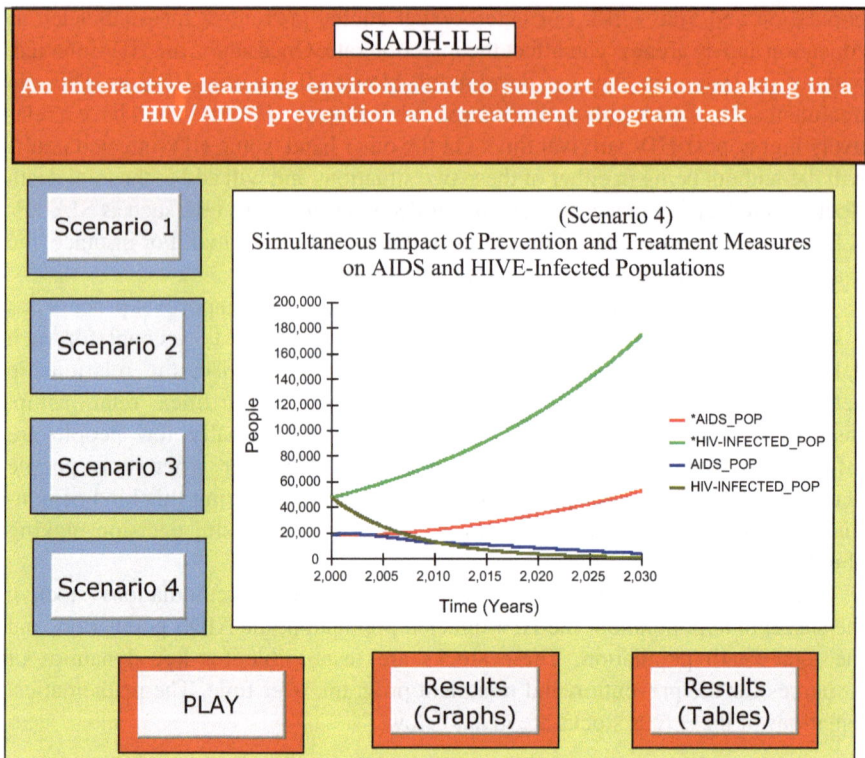

Fig. 3.3 The decision panel of SIADH-ILE. (Source: Qudrat-Ullah and Tsasis 2017)

of their decisions in both a graphical and tabular format. For instance, in Fig. 3.3, the results of Scenario 4 are displayed graphically.

3.3.1   Model Validation

System dynamics models are causal models, well suited for policy analysis and assessment rather than the point prediction of variables under study. Therefore, the validation of these models is the process of building confidence in the useful-ness of a model (Forrester 1961; Qudrat-Ullah and BaekSeo 2010; Roberts and Dangerfield 1992). Both the structural and behavior validity procedures are applied to system dynamics models. While structural validity ascertains that model structures (e.g., feedback loops) generate the right behavior, behavior validity assesses how well the model-generated behavior matches the observed patterns of the real system (Qudrat-Ullah 2008). Our SIADH model was success-fully evaluated against all these procedures (for details, please see in Qudrat-Ullah and BaekSeo (2010)).

3.4   Procedures and Participants

The task program (SIADH-ILE) was installed and pilot[2]-tested prior to the main study. All subjects were supplied with a folder containing the consent form, instructions to lead them through a session, training materials for the task, note-pads, and pens as they were encouraged to take notes during the experiment. The experiment started with each participant returning the signed consent form and taking a pretest on structural and heuristic knowledge. Then the experimenter introduced the task system and the experiment. All of the groups received the same general instructions by the same facilitator. To avoid novelty and the Hawthorne effects, all subjects completed a training trial, making decisions in each period, accessing and observing feedback on their decisions via graphs and tables. In this training trial, a simplified version of the dynamic task, where delays and nonlinear relationships were turned off, was used. Then, all subjects completed two formal trials interceded by either a small break for the control group or a debriefing activ-ity for the experimental groups.

We conducted our experiment with 96 executive MBA program participants, recruited from three Canadian universities. They were randomly assigned to each of the three groups.

Group 1: With process-oriented debriefing (PD)
Group 2: With outcome-oriented debriefing (OD)
Group 3: No debriefing (ND)

The use of business graduate students as surrogates for practicing managers is considered adequate because SIADH-ILE task is an information processing-oriented task and the information processing behavior of students and managers rarely differs in dynamic tasks (Sing 1998; Ford and Mccormack 2000). Besides, all of our participants were working adults. Their work experience ranged from 3 to 5 years. The nature of experience ranged from senior supervisory level to middle level of management. These working students were highly motivated and eager to improve their decision making skills to enhance their prospects for better jobs, pro-motions, and salary packages (Qudrat-Ullah 2007).

Subjects were told that scores on the practice trial would not count. In addition, subjects in each of the debriefing conditions were given unique information through separate sessions. The experiment was conducted on desktop computers in three computer labs, each with a seating capacity of 50. At most, 32 subjects participated in a session. Sessions were staggered. On arrival, each of the participants was seated at a randomly assigned desktop computer, signed a consent form, received general instructions on the tasks, completed a pre-task knowledge test and demographic questionnaire, completed the practice and two main trials interceded by the respec-tive form of debriefing, and completed the knowledge and debriefing manipulation check questionnaires (Table 3.2).

---

[2]The pilot test was conducted with nine subjects. The performance data is not included in this study.

Table 3.2 The experimental procedures

Step	Activity
1	*Greeting and introducing the steps below* Greeted by the investigator Signed the consent form Asked to go through the game description material to develop the understanding of the task Encouraged to ask questions, if any Assigned to an available computer Answered the pre-task knowledge questionnaire (Appendix C) Assigned with an anonymous ID indicating the treatment group Reassigned to an available computer (in case of dyads)
2	*Training session* Went through the computer screens for the training session Had been told in Step 1 that the same materials were also available in the folder
3	*One practice trial and two formal trials* Directed to the practice trial at the last screen of the training session Told that score for practice trial not counted Started the two formal trials following the practice trial
4	*Post-task debriefing* Directed to the post-task debriefing when the last trial was done Accomplished the post-task questionnaire (Appendix D) Instructed to inform the investigator when done Left the classroom with a note to report again after 2 weeks for the final stage of the experiment and about the payment (if any)
5	*Post-game debriefing* Greeted by the investigator Accomplished the post-gaming questionnaire (retention test: Appendix F) Informed about the top performers Directed to collect the payment Left the room with a note about further contact information

3.5   Decision Making and Task Performance in SIADH-ILE

In the context of limited resources and competing priorities, the development and use of scenarios help managers to make better decisions (Adobor and Daneshfar 2006; Lane 1995). The learning objective of the users of the scenario-based SIADH-ILE is to "craft a policy scenario that produces fewer HIV/AIDS-related deaths at a relatively lower cost." In the context of treatment, the HAART vaccine is the only option where users decide how much to spend on improving its efficacy. More R & D investments result in increased average, efficacious time. On the other hand, for preventive care, two measures related to the acquisition and spread of heterosexual infections, (i) contact frequency and (ii) infectivity, are considered. Here, the users make decisions on how much to spend on education awareness initiatives to improve preventive care (by a reduction in both the contact frequency and infectivity). Details on these measures are given in Qudrat-Ullah (2014). Learners practice with the following four scenarios at their disposal and then design their best policy scenario. Their performance is measured through the count of HIV/AIDS-related deaths and cost (i.e., minimum deaths with the least cost prevention and treatment care is the best performance) over the course of 30 years. The four scenarios are:

Scenario 1: (Baseline) This scenario represents the business-as-usual case.

Scenario 2: This scenario assumes that HAART treatment is available and the HAART is efficacious for an average of **x** years (users specify the years from 10 to 20 and observe the cost and lives saved).

Scenario 3: Here, a combined improvement of **x** % both in the contact frequency and the infectivity is assumed (users specify these two measures (in a range of 40–80%) and observe the cost and lives saved).

Scenario 4: In this scenario, the simultaneous impact of prevention (i.e., Scenario 3) and treatment (i.e., Scenario 2) measures is considered.

The users of SIADH-ILE have to play these three scenarios before making their decisions (i.e., how many years of HAART efficacy, how much % improvement in contact frequency, and how much % improvement in the infectivity) in Trials 1 and 2. All subjects completed two main trials. Subjects took between 110 and 120 minutes in total to complete the experiment.

3.6 Design of Debriefing Sessions

The independent variable is the "availability of debriefing" in an SDILE. To avoid any facilitator or debrief effects, all the debriefing sessions were delivered by a single facilitator. Debriefing sessions were designed based on the Lederman (1992) model that includes three phases:

(i) The introduction to the systematic reflection and analysis-based debriefing
(ii) Analysis of subjects' performance and experiences with the dynamic task
(iii) The generalization and relevance to real-world decision making

In the process-oriented debriefing, subjects' performance charts were shown and discussed to relate the structure of the system with its behavior (Davidsen and Spector 1997). Outcome-oriented debriefing focused on a "benchmark" solution. Each session took 30 minutes. Table 3.3 lists the key activities of both debriefing sessions.

3.7 Measurements

The dependent variables are decision strategy, decision time, task performance, structural knowledge, and heuristic knowledge. The task performance metric is chosen so as to assess how well each subject did relative to a benchmark rule

Table 3.3 Key activities of debriefing sessions

Outcome-oriented debriefing	Process-oriented debriefing
Goal and objectives of the session and the dynamic task are clarified	Goal and objectives of the session and the dynamic task are clarified
Performance/decision outcome charts (three anonymous charts are randomly picked) are displayed	Performance/decision outcome charts (three anonymous charts are randomly picked) are displayed
Benchmark solution is presented	Structure-behavior graphs are presented
Discussions with a focus on the gap between the benchmark solution and their solutions is facilitated	The discussion focuses on causal relationships between task variables and performance outcomes

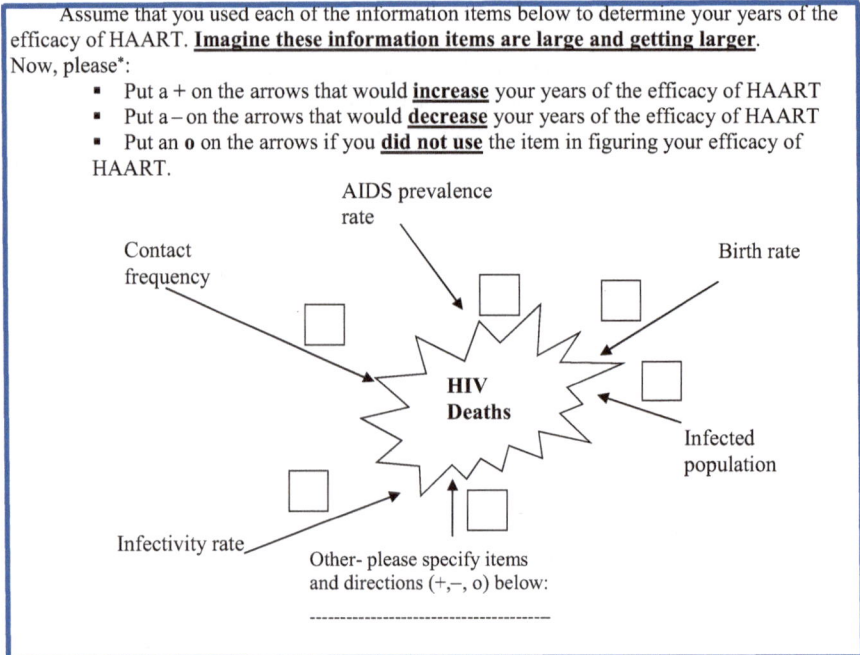

Assume that you used each of the information items below to determine your years of the efficacy of HAART. **Imagine these information items are large and getting larger**. Now, please*:

- Put a + on the arrows that would **increase** your years of the efficacy of HAART
- Put a – on the arrows that would **decrease** your years of the efficacy of HAART
- Put an **o** on the arrows if you **did not use** the item in figuring your efficacy of HAART.

AIDS prevalence rate

Contact frequency

Birth rate

HIV Deaths

Infected population

Infectivity rate

Other- please specify items and directions (+,–, o) below:

Fig. 3.4 Example of heuristic knowledge test questions. ∗An irrelevant example was used to explain how to accomplish this question

(Qudrat-Ullah 2014). A posttest questionnaire measured the structural knowledge through 14 closed-ended questions on the relationships between pairs of the task variables. For example, Other things being equal, an increase in years of the efficacy of HAART leads to (1 *immediate increase*, 2 *delayed increase*, 3 *no change*, 4 *immediate decrease*, 5 *delayed decrease [subjects were asked to choose one of the five answers]*) HIV deaths. The heuristic knowledge of the dynamic interrelationships across decisions was assessed through the use of causal loop diagrams. Figure 3.4 presents an example of heuristic knowledge test questions.

The decision time was measured as the time spent by a subject making decisions in each of the decision periods. It is also used as a surrogate to "effort" in this study. SIADH-ILE is designed to record the decision time for all 30 periods. SIADH-ILE was programmed to use a consistency metric (Sengupta and Abdel-Hamid 1993) to generate data for subjects' decision strategies. To account for the nature of fluctuations in the decision strategies, two measures, the average amount of fluctuations and the number of reversals in the direction of the decisions, are used. For example, if a decision maker increased the efficacy of HAART as 3, 4, 2, and 3 years consecutively, then the average amount of fluctuations in his decisions would be the absolute value of $[(-1 + 2 + 1)/4] = 0.5$, and the number of reversals would be 3. Both the frequency and the magnitude of fluctuations will allow us to classify decision makers' strategies as SVS or not.

To overcome the traditional barriers of academic-practitioner knowledge transfer (Barnett and Ceci 2002), all participants were asked to come back after 6 weeks, during the same semester, to participate in another SDILE: LichenBankILE (Qudrat-Ullah 2014) for transfer learning assessment. The choice of this elapsed time of 6 weeks is based on the Barnett taxonomy of far transfer (Barnett and Ceci 2002). The subjects' performance in this task will be used to assess their transfer learning. The subjects' transfer learning was measured using a five-point Likert scale-based questionnaire consisting of 13 questions about a different dynamic task, LichenBankILE.[3] Each question is about the causal link between two variables of the task system. For instance, an increase in Lichen Density leads to ……………………… in Feed per Reindeer:

(a) ………… immediate increase
(b) ………… delayed increase
(c) ………… no change
(d) ………… immediate decrease
(e) ………… delayed decrease.

3.8 Summary of the Experimental Procedures and Protocols

To test our proposed set of hypotheses, we adopted an experimental approach which allowed us to apply several select controls. We began with the description of our research design which provides justification for the various procedures that we applied to collect and analyze the data. The description of background education and the experience of subjects was our next step. This allowed us to see if both the experimental and our control groups were homogenous or not. The dynamic task, SIADH-ILE, is described in detail: its macro-level view of key structural elements

---

[3]This questionnaire is the short version of a questionnaire which was published in Qudrat-Ullah (2014).

like feedback and delays, its mathematical equations, its decision panel, and its help systems. Also, the validity testing of the simulation model, SIADH, is described. Design and implementation of debriefing both in terms of process-oriented and outcome-oriented debriefing are explained. How we measure the SIADH-ILE users' decision making and learning on five measures (i.e., task performance, decision strategy, decision time, structural knowledge, and heuristic knowledge) is illustrated. Also, the overall protocol used in our laboratory experiments is explicitly laid down. In the next chapter, we present the key results that we obtained from these laboratory experiments.

Appendix A: Transfer Learning Questionnaire

In the following, you are asked to reflect on your understanding about LichenBankILE game by rating the relationships between variables in the underlying model of the game? In each of the following questions, imagine that other things being equal, left-hand bold face variable below either increases or decreases. What will happen to the right variable? Will it increase immediately or after a delay? Will it remain unchanged? Will it decrease immediately or after a delay?

An immediate increase implies that an increase or decrease in the left variable is followed by an increase in the right variable in the same year. No change means that if the left variable stays constant after an initial increase or decrease, so will be the right variable. In a delayed relationship, however, the right variable continues to increase or decrease for some time even after the left variable is constant (of course initially it was either increased or decreased). As an example, if the left variable is "Hiring New Employees" and the right variable is "Retirements," other things being equal, an increase in "Hiring New Employees" leads to a delayed increase in "Retirements," because "Retirements" continue to increase even after "Hiring New Employees" has stopped. New Employees remain in the pipeline for some time (hopefully till their retirement age!), and "Retirements" thus are a delayed function of "Hiring New Employees."

You make choices by clicking on the appropriate button. Having done that, you have about 25 seconds to make the next choice. A timer will appear in the upper right corner of this screen once you have finished this introduction.

Here is the set of questions that will appear on your screen:

1. An increase in *Herd Recruitment* leads to .... in *Lichen Stock*

    (a) .... immediate increase
    (b) .... delayed increase
    (c) .... no change
    (d) .... immediate decrease
    (e) .... delayed decrease

2. An increase in *Herd Stock* leads to .... in *Lichen Harvesting*

   (a) .... immediate increase
   (b) .... delayed increase
   (c) .... no change
   (d) .... immediate decrease
   (e) .... delayed decrease

3. An increase in *Herd Stock* leads to .... in *Lichen Density*

   (a) .... immediate increase
   (b) .... delayed increase
   (c) .... no change
   (d) .... immediate decrease
   (e) .... delayed decrease

4. An increase in *Lichen Density* leads to .... in *Feed per Reindeer*

   (a) .... immediate increase
   (b) .... delayed increase
   (c) .... no change
   (d) .... immediate decrease
   (e) .... delayed decrease

5. An increase in *Feed per Reindeer* leads to .... in *Lichen Harvesting*

   (a) .... immediate increase
   (b) .... delayed increase
   (c) .... no change
   (d) .... immediate decrease
   (e) .... delayed decrease

6. An increase in *Lichen Stock* leads to .... in *Reindeer Birth Rate*

   (a) .... immediate increase
   (b) .... delayed increase
   (c) .... no change
   (d) .... immediate decrease
   (e) .... delayed decrease

7. A decrease in Herd Stock leads to .... in Natural Death Rate of Reindeer

   (a) .... immediate increase
   (b) .... delayed increase
   (c) .... no change
   (d) .... immediate decrease
   (e) .... delayed decrease

8. A decrease in *Lichen Harvesting* leads to .... in *Lichen Stock*

    (a) .... immediate increase
    (b) .... delayed increase
    (c) .... no change
    (d) .... immediate decrease
    (e) .... delayed decrease

9. A decrease in *Herd Stock* leads to .... in *Operating Costs*

    (a) .... immediate increase
    (b) .... delayed increase
    (c) .... no change
    (d) .... immediate decrease
    (e) .... delayed decrease

10. A decrease in *Lichen Density* leads to .... in *Operating Costs*

    (a) .... immediate increase
    (b) .... delayed increase
    (c) .... no change
    (d) .... immediate decrease
    (e) .... delayed decrease

11. A decrease in *Operating Cost* leads to .... in *Operating Profits*

    (a) .... immediate increase
    (b) .... delayed increase
    (c) .... no change
    (d) .... immediate decrease
    (e) .... delayed decrease

12. A decrease in *Weight per Reindeer* leads to .... in *Revenue*

    (a) .... immediate increase
    (b) .... delayed increase
    (c) .... no change
    (d) .... immediate decrease
    (e) .... delayed decrease

13. A decrease in *Herd Stock* leads to .... in *Revenue*

    (a) .... immediate increase
    (b) .... delayed increase
    (c) .... no change
    (d) .... immediate decrease
    (e) .... delayed decrease

References

Adobor, H., & Daneshfar, A. (2006). Management simulations: Determining their effectiveness. *Journal of Management Development, 25*(2), 151–168.

Atun, R. A., & Sittampalam, S. (2006). A review of the characteristics and benefits of SMS in delivering healthcare. In R. A. Atun (Ed.), *The role of mobile phones in increasing accessibility and efficiency in healthcare.* Vodafone Group Plc.

Barnett, S., & Ceci, S. (2002). When and where do we apply what we learn? A taxonomy for far transfer. *Psychological Bulletin, 128*(4), 612–637.

Davidsen, P. I., & Spector, J. M. (1997). Cognitive complexity in system dynamics based learning environments. In *International system dynamics conference* (pp. 757–760). Istanbul: Bogazici University Printing Office.

Ford, D. N., & Mccormack, D. E. M. (2000). Effects of time scale focus on system understanding in decision support systems. *Simulation and Gaming, 31*(3), 309–330.

Forrester, J. W. (1961). *Industrial dynamics.* Cambridge, MA: Productivity Press.

Homer, J. B., & Hirsch, G. B. (2006). System dynamics modeling for public health: Background and opportunities. *American Journal of Public Health, 96*(3), 452–458.

Lane, D. C. (1995). On a resurgence of management simulations and games. *Journal of the Operational Research Society, 46*, 604–625.

Lederman, L. C. (1992). Debriefing: Towards a systematic assessment of theory and practice. *Simulation and Gaming, 23*(2), 145–160.

Moxnes, E. (2004). Misperceptions of basic dynamics: The case of renewable resource management. *System Dynamics Review, 20*, 139–162.

Qudrat-Ullah, H. (2007). Debriefing can reduce misperceptions of feedback hypothesis: An empirical study. *Simulations and Gaming, 38*(3), 382–397.

Qudrat-Ullah, H. (2008). Behavior validity of a simulation model for sustainable development. *International Journal of Management and Decision Making, 9*(2), 129–139.

Qudrat-Ullah, H. (2010). Perceptions of the effectiveness of system dynamics-based interactive learning environments: An empirical study. *Computers and Education, 55*, 1277–1286.

Qudrat-Ullah, H. (2014). Yes we can: Improving performance in dynamic tasks. *Decision Support Systems, 61*, 23–33.

Qudrat-Ullah, H., & BaekSeo, S. (2010). How to do structural validity of a system dynamics type simulation model: The case of an energy policy model. *Energy Policy, 38*(5), 2216–2224.

Qudrat-Ullah, H., Saleh, M. M., & Bahaa, E. A. (1997). Fish Bank ILE: An interactive learning laboratory to improve understanding of 'The Tragedy of Commons'; a common behavior of complex dynamic systems. *Proceedings of 15th international system dynamics conference,* Istanbul, Turkey.

Qudrat-Ullah, H., & Tsasis, P. (Eds.). (2017). *Innovative healthcare systems for the 21st century.* Cham: Springer. ISBN 978-3-319-55773-1.

Roberts, C. A., & Dangerfield, B. C. (1992). Estimating the parameters of an AIDS spread model using optimization software: Results for two countries compared. In J. A. M. Vennix, J. Faber, W. J. Scheper, & C. A. T. Takkenberg (Eds.), *System dynamics 1992* (pp. 605–617). Cambridge, MA: System Dynamics Society.

Sengupta, K., & Abdel-Hamid, T. (1993). Alternative concepts of feedback in dynamic decision environments: An experimental investigation. *Management Science, 39*, 411–428.

Sing, D. T. (1998). Incorporating cognitive aids into decision support systems: The case of the strategy execution process. *Decision Support Systems, 24*, 145–163.

Spector, J. M. (2000). System dynamics and interactive learning environments: Lessons learned and implications for the future. *Simulation and Gaming, 31*(4), 528–535.

Tebbens, D., & Thompson, K. (2009). Priority shift- ing and the dynamics of managing eradicable infectious diseases. *Management Science, 55*, 650–663. https://doi.org/10.1287/mnsc.1080.0965.

Chapter 4
Results of Experimental Research

Abstract In this chapter, we present our key results. Specifically, our stated hypotheses are explicitly tested here where the efficacy of debriefing-based SDILE is assessed on five dimensions: subjects' task performance, decision strategy, decision time, structural knowledge, and heuristic knowledge. Both the main effects and indirect effects of debriefing on these five dimensions are reported. Beginning with the descriptive analysis of our participant, we report the subjects' performance and learning in the dynamic task. Subjects' reaction to received debriefing as well as the effects of the practice is also reported here.

Keywords SDILE · Task performance · Decision strategy · Decision time · Heuristic knowledge · Debriefing · Effects of the practice · Dynamic tasks · System dynamics · Health policy · HAART · Multivariate analysis · Process-oriented debriefing · Outcome-oriented debriefing

4.1 Introduction

Successful decision making in dynamic tasks is a much sought after expertise. Organizations want tools, techniques, ideas, and concepts that can help them have a skilled workforce capable enough to deal with the uncertainties and complexities of the dynamic task. In an effort to address this ongoing managerial need, we set the objective for this book to cope up with a viable and cost-effective solution. In this chapter, we present the results of the experimental research we conducted for this book project.

As with any evaluative study, first we describe who our participants were and what was their background education and work experience. To understand the comparative effectiveness of our two interventions, process-oriented debriefing and outcome-oriented debriefing, it was critical to have as homogenous as possible participants for our experimental groups. Task performance of each group with a randomly selected user's profile of actual decision making is presented. As the subjects' decision strategies influence their performance, we used specific metrics (Abdel-Hamid et al. 1999) to calculate the number of fluctuations as well as the number of reversals in their decision strategies. Decision efficiency is measured as the total time a subject spent in making a decision on each of 30 periods of the simulation task, SIADH-ILE. Then the results pertaining to the subjects' learning, structural knowledge and heuristic knowledge, are shown.

4.2   Our Decision Makers and Learners

Table 4.1 presents the demographics, background education, and prior task knowledge of our decision makers and learners. There were no significant differences across treatments with respect to gender ($p = 0.872$ both for males and females), age ($F = 0.31$, $p = 0.567$), prior structural knowledge ($F = 0.9$, $p = 0.871$), and heuristic knowledge ($F = 0.73$, $p = 0.342$) about the task system. Likewise, all the groups did not differ in terms of their background education. The majority of the participants in each of the three groups had taken statistics, accounting, operations management, and health policy courses. None of the subjects had taken system dynamics-related courses. Therefore, it is safe to assume that even if these demographic factors had

Table 4.1 Participant demographics and background education and knowledge

Variable	ND group ($n = 32$)	OD group ($n = 32$)	PD group ($n = 32$)	Test statistic	Critical value	p[a]-value
Age	34.2 (1.1)	33.7 (1.4)	33.2 (169)	$F = 0.31$	3.29	0.567
Gender						
Male	49%	49%	48%	$\chi^2 = 0.97$	26.12	0.872
Female	51%	51%	52%	$\chi^2 = 3.97$	30.37	0.872
Academic background						
FAME	1.3	1.2	1.3	$F = 0.03$	3.29	0.823
CIMS	0.6	0.5	0.7	$F = 0.04$	3.29	0.782
HPR	1.2	1.1	1.3	$F = 0.05$	3.29	0.667
Structural knowledge	13.7 (2.0)	13.7 (2.1)	14.0 (17.8)	$F = 0.9$	3.29	0.871
Heuristic knowledge	4.09 (0.75)	4.10 (0.77)	4.01 (0.59)	$F = 0.73$	3.29	0.342

Values in parentheses are standard deviations. FAME = finance, accounting, management, and economics (score = 1 if course taken, score = 0 if course not taken; maximum = 4, minimum = 0); CIMS = computer science, IT, mathematical modeling, and system dynamics (maximum = 4, minimum = 0); HPR= health policy-related subjects (maximum = 4, minimum = 0). The average score on the structural knowledge test (maximum score on structural test = 25, minimum score = 0) and the average score on heuristic knowledge test (maximum score on heuristic test = 15, minimum score = 0) are given
[a]Alpha value is 0.05 unless stated otherwise

Table 4.2 Between subject effects of debriefing

Dependent variable	F-statistic	p-value
Decision strategy (AAF)	F(2.93) = 31.563	0.000
Decision strategy (REV)	F(2.93) = 22.02	0.000
Decision time	F(2.93) = 17.21	0.000
Task performance	F(2.93) = 45.13	0.000
Structural knowledge	F(2.93) = 212.32	0.000
Heuristic knowledge	F(2.93) = 217.78	0.000

AAF average amount of fluctuations, *REV* reversals in the direction of decisions

any effect on the participants' performance, the effect was same for the three groups and did not materially affect the results of this study.

4.3 Main Effects of Debriefing

Debriefing appears to have a significant effect on subject decision making and learning in a dynamic task. There was a significant difference among the three groups when considered jointly on the variables of decision processes (i.e., decision strategy and decision time) and decision outcomes (task performance, structural knowledge, and heuristic knowledge), Pillai's trace = 1.13, $F = 29.49$, $p = 0.000$, partial $\eta^2 = 56$. A separate ANOVA for each of the dependent variables also confirmed the existence of significant differences among the treatment groups on the type of debriefing (Table 4.2).

4.4 Debriefing and Task Performance

Table 4.3 presents the results of several ANOVAs for the main effects of all the levels of debriefing on subjects' performance. We conducted a planned contrast analysis among all the three groups on all the dependent variables. Compared with the group with no debriefing (ND), all the treatment groups performed significantly better on task performance, so the hypothesis H1 is strongly ($p = 0.000$) supported. The means of the OD group were significantly lower than the PD group ($p = 0.000$). Thus, subjects receiving process-oriented outperformed the outcome-oriented group. Hypothesis H1a, therefore, is supported. It is worth mentioning that the group with process-oriented debriefing performed the best but still did not beat the benchmark[1] (BM) performance (i.e., achieved the cumulative deaths about 11% more the benchmark's HIV deaths over a period of 30 years).

---

[1] The benchmark performance is based on a near optimal strategy which is described in detail in Qudrat-Ullah (2014).

Table 4.3 Comparative effects of debriefing on decision making and learning

Dependent variable	Groups (mean, standard deviation)		Significance	Hypotheses
Decision strategy (amount of fluctuations)	FD HD	ND (3.11, 0.21) vs. OD (2.04, 0.14) ND (3.12, 0.26) vs. OD (2.19, 0.12)	$p = 0.000$ $p = 0.000$	H2 and H2a are supported
	FD HD	ND (3.11, 0.21) vs. PD (0.45, 0.11) ND (3.14, 0.23) vs. PD (0.22, 0.13)	$p = 0.000$ $p = 0.000$	
	FD HD	OD (2.03, 0.21) vs. PD (0.45, 0.14) OD (2.17, 0.19) vs. PD (0.21, 0.12)	$p = 0.000$ $p = 0.000$	
Decision strategy (# of reversals)	FD HD	ND (17, 2.31) vs. OD (11, 2.03) ND (18, 2.31) vs. OD (11, 2.11)	$p = 0.000$ $p = 0.000$	
	FD HD	ND (13, 2.32) vs. PD (4, 1.66) ND (14, 2.48) vs. PD (3, 1.02)	$p = 0.000$ $p = 0.000$	
	FD HD	OD (11, 2.03) vs. PD (4, 1.66) OD (11, 2.11) vs. PD (3, 1.02)	$p = 0.000$ $p = 0.000$	
Decision time	ND (36.92, 7.83) vs. OD (22.91, 5.55)		$p = 0.000$	H3 is supported but H3a is not supported
	ND (36.92, 7.83) vs. PD (38.32, 4.12)		$p = 0.010$	
	OD (22.91, 5.55) vs. PD (38.32, 4.12)		$p = 0.000$	
Structural knowledge	ND (14.10, 1.56) vs. OD (18.72, 2.21)		$p = 0.000$	H4 and H4a are supported
	ND (14.10, 1.56) vs. PD (23.98, 1.65)		$p = 0.000$	
	OD (18.72, 2.21) vs. PD (23.98, 1.65)		$p = 0.000$	

(continued)

Table 4.3 (continued)

Dependent variable	Groups (mean, standard deviation)	Significance	Hypotheses
Heuristic knowledge	ND (3. 98, 1.22) vs. OD (7.12, 1.21)	$p = 0.000$	H5 and H5a are supported
	ND (3. 98, 1.22) vs. PD (14.11, 1.64)	$p = 0.000$	
	OD (7.12, 1.21) vs. PD (14.11, 1.64)	$p = 0.000$	

ND group with no debriefing, *OD* group with outcome-oriented debriefing, *PD* group with process-oriented debriefing, *FD* contact frequency decisions, *HD* HAART effectiveness decisions; task performance (TP) is the % (percent) deviation from the benchmark in Trial 2. A score of zero means that a participant's performance did not reach the level of the benchmark. A score of greater than zero indicates that the participant has performed better than the benchmark. A score of less than zero indicates that the participant has performed poorer than the benchmark

Figure 4.1 displays the task performance, measured in some AIDS-related deaths, of the User-M (randomly selected). Compared with the average performance of base-case scenario (business-as-usual case), this user performs very well (F-value = 29.2; p-value = 0.000). In fact, if User-M's policy is implemented, cumulatively there are 23,205 lives saved throughout 30 years. When we looked at the decisions of the User-M that lead to achieving this performance, a 12-year efficacy of HAART, a 51% improvement in contact frequency, and a 62% improvement in the infectivity (i.e., the rate of infection) through investments in technology, education, and awareness programs across Canada were the drivers. We also looked at several other users' task performance and found similarity in their decisions (i.e., the selections of the parameters of the dynamic task). Training with scenario-based SIADH-ILE appears to support the subject's decision making and learning in the dynamic task.

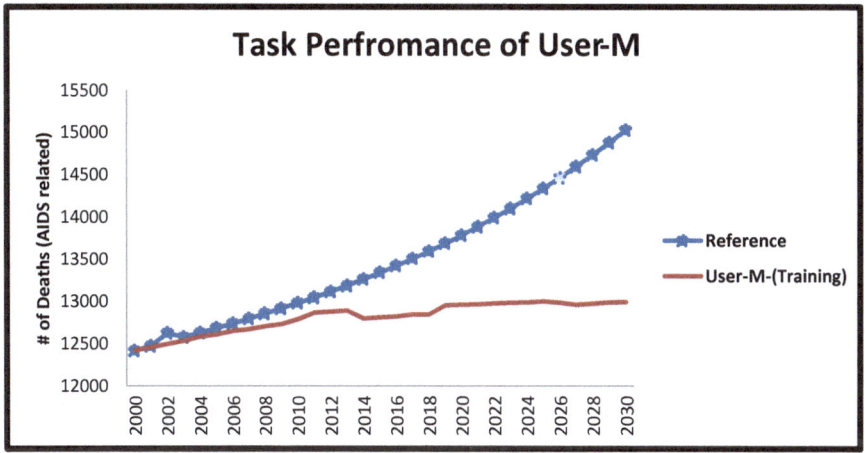

Fig. 4.1 Task performance improvement (of User-M) in SIADH-ILE

4.5   Debriefing and Decision Strategy

Decision strategies shape the subject's decisions and offer some useful insights into their decision making process. Having had access to useful causal information on contact frequency decision and HAART effectiveness decision through a structured discussion on structure-behavior graphs of the dynamic task, these subjects managed the task well. The average amount of fluctuations, in the exploratory-type decision strategies of the subjects in PD group, was 0.46, and the number of reversals was 3 (see Table 4.4). The pattern for the OD group lies in between the two trends. In accordance with the expert solution-based information given to them, these subjects built up and use their fleet early. However, the fluctuations in their decisions are higher than that of the PD subjects, apparently because of their inability to diagnose system changes and build causal understating about the structure-behavior relationships of the variables of the task.

Table 4.4 summarizes the overall fluctuations in the subjects' decision patterns. Subjects in the ND group fluctuated the most (and most often) in their decisions, followed by those in the OD and PD groups, respectively, regardless of the trials. A multivariate analysis of the two variables representing fluctuation indicates a significant main effect for type of debriefing ($p = 0.000$), indicating that the fluctuation in subjects' decisions was contingent on the type of debriefing they were given. Posterior tests also show that the PD group and OD group had significantly less fluctuation than the ND group on both dependent variables. The fluctuation in the PD group was significantly less than the OD group with respect to the amount of fluctuation, as well as the number of reversals in the direction both in the case of ship ordering and ship utilizing decisions—subjects in PD group employed SVS more than those in OD group. Therefore, our hypotheses H2 and H2a stand supported.

Table 4.4   Fluctuations in subjects' decisions (means (M))

Groups	Contact frequency decisions (Trial 1 vs. Trial 2)	HAART effectiveness decisions (Trial 1 vs. Trial 2)
(a) Average amount of fluctuations		
ND	3.15 vs. 3.19	3.14 vs. 3.09
OD	2.23 vs. 2.02	2.63 vs. 2.15
PD	2.71 vs. 0.46	2.76 vs. 0.22
(b) Reversals in direction of fluctuations.		
ND	15 vs. 14	13 vs. 14
OD	11 vs. 8	11 vs.8
PD	12 vs. 3	12 vs. 3

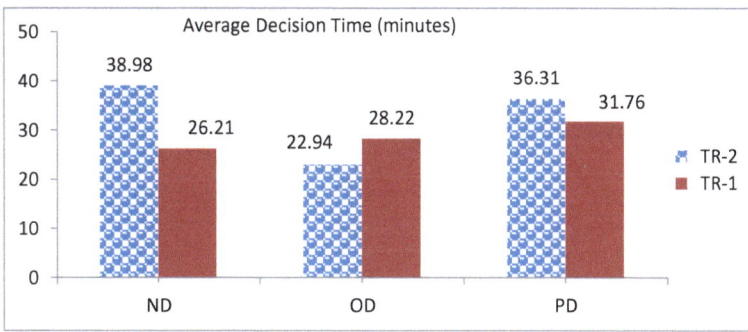

Fig. 4.2 Effects of debriefing on decision time

4.6   Debriefing and Decision Time

Table 4.3 shows that the effects of debriefing on subjects' decision time are mixed. The hypothesis H3 that both OD group and PD group will be more efficient than the ND group is supported, but H3a is not supported. Contrary to the hypothesis, H3a, *users of an SDILE with process-oriented debriefing will become more efficient deci-sion makers than users of an SDILE with outcome-oriented debrefing*, the group with a process-oriented debriefing where the subjects were expected to spend less time due to the availability of structure-behavior elucidating causal information spent significantly more ($p = 0.000$) time ($M = 36.3$) than those with outcome-ori-ented debriefing ($M = 22.9$). Overall, ND group spent the longest time of about 39 minutes in their Trial 2 (Fig. 4.2). In terms of comparative decision time across the two trials, only those in group OD spent less time in Trial 2 than in Trial 1. It appears that subjects in the OD group made use of expert solution-based heuristics rather than adapting to an exploratory type of strategy to make their decisions. However, both PD and ND groups spent more time in Trial 2 than in Trial 1. A simultaneous look at the decision patterns of these two groups in Fig. 4.2 shows that decision makers with process-oriented debriefing adapted to a systematic explor-atory strategy that takes times while those in ND group perhaps more time in figur-ing out causal relationships between task variables.

4.7   Debriefing and Structural Knowledge

Table 4.3 presents the results for the main effects of all the levels of debriefing on structural knowledge. Compared with the group with no debriefing, all the treat-ment groups performed significantly better ($p = 0.00$). Therefore, the hypotheses H4, *subjects receiving any form of debriefing (OD or PD) will outperform those without any debriefing (ND) on structural knowledge*, and H4a, *subjects receiving process-oriented debriefing will outperform those who receive outcome-oriented*

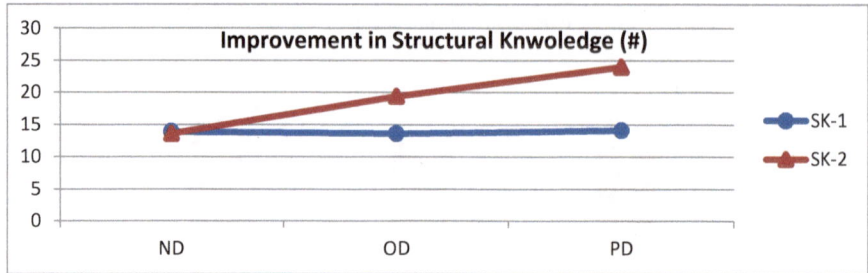

Fig. 4.3 Structural knowledge improvement across the trials. **SK1** structural knowledge in Trial 1, **SK-2** structural knowledge in Trial 2

debriefing on structural knowledge, are strongly supported. Overall, the "process-oriented debriefing" group performed the best on structural knowledge (24), and "no debriefing" group scored 13.60. The OD group scored 19. Across the trials, PD group appears to make use of available causal information about the variables of the task system and performed the best (Fig. 4.3): subjects in PD group showed an improvement of over 70%. ND group, on the other hand, exhibited a decline of 2%.

4.8   Debriefing and Heuristic Knowledge

Table 4.3 presents the results for the main effects of all the levels of debriefing on heuristic knowledge. Compared with the group with no debriefing, all the treatment groups performed significantly better ($p = 0.000$). Therefore, the hypotheses H5, *subjects receiving any form of debriefing (OD or PD) will outperform those without any debriefing (ND) on heuristic knowledge, and* H5a, *subjects receiving process-oriented debriefing will outperform those who receive outcome-oriented debriefing on heuristic knowledge,* are strongly supported. Overall, again, among the three groups, the "process-oriented debriefing" group performed the best on structural

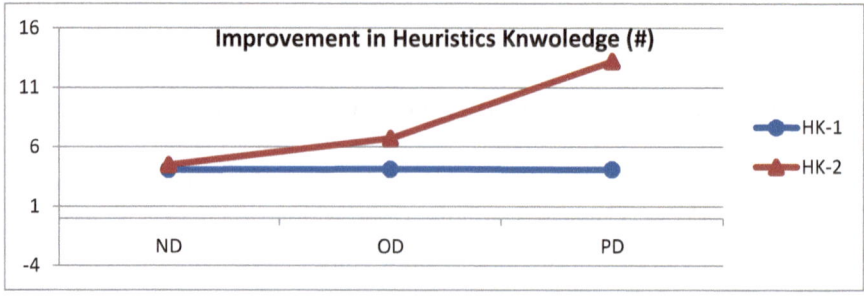

Fig. 4.4 Heuristic knowledge improvement across the Trials. **HK1** heuristic knowledge in Trial1, **HK-2** heuristic knowledge in Trial 2

knowledge (13.21 out of 15), and "no debriefing" group scored the lowest (4.48 out of 15). The OD group scored 6.72 out of 15. Across the trials, the PD group performed the best (Fig. 4.4): subjects in the PD group showed an impressive improvement of over 200%. ND group, on the other hand, exhibited a little improvement of about 9%. Subjects in the OD group performed in between with an improvement of 63%.

4.9   Perceived Effects of Debriefing

Perceptions make a reality. To test whether the two experimental groups that received debriefing differed in their reactions to debriefing, a MANOVA was conducted using the two self-reported scales (usefulness of debriefing in arousing interest in the task (DIT) and usefulness of debriefing in understanding the task (DUT)) as dependent measures. The MANOVA results indicated that treatment groups differed on the two measures, Wilk's $\lambda = 0.02$, F (1, 65) = 213, $p = 0.00$.

Based on separate ANOVAs, significant effects for the debriefing manipulation were found for DIT (F (4, 62) = 14.21, $p = 0.00$) and DUT (F (4, 62) = 11.45, $p = 0.00$). Post hoc comparisons indicated that DIT was significantly lower in the OD group ($M = 2.45$) than of the PD group. The subjects in the PD group rated the usefulness of debriefing in arousing interest in the task very high ($M > = 4$). On the perceived utility of the debriefing, DUT, OD group rated it low ($M = 2.66$). Like in the case of DIT, the PD group rated the usefulness of debriefing in understanding the task very high ($M > = 4.41$ out of 5).

4.10   Subjects' Performance Across the Trials: Effects of Practice or Effects of Debriefing

The performance improvement from Trial 1 to Trial 2 for all the groups was also analyzed. The group with no debriefing did not show any improvement in task performance. On the other hand, both the process-oriented debriefing and outcome-oriented debriefing showed significant improvement. Table 4.3 presents similar results on subjects' use of systematic variable strategies. On decision time, the only group with outcome-oriented debriefing used less time in Trial 2 than in Trial 1. Again on structural knowledge and heuristic knowledge group with no debriefing did show any improvement. On the other hand, both groups with debriefing showed a significant improvement in their knowledge. A simultaneous look at these all results suggests that it would be naïve to attribute the significant performance improvement on task performance, decision strategy, structural knowledge, and heuristic knowledge by the groups with debriefing to a practice effect alone. Thus, there appears no real threat to the established support of the hypotheses we presented in earlier sections.

4.11 Summary of the Main Results

With the objective of improving people's decision making and learning in dynamic tasks, we proposed and tested several hypotheses. Overall, we found significant support for the effectiveness of debriefing-based SDILE. In particular, process-oriented debriefing allowed the subjects to improve their task performance, adapt consistent decision strategies, and develop better heuristic and structural knowledge about the task system. Consistent with the logic of "teach them how to fish rather than giving them a fish," process-oriented debriefing, although took more time, made the decision makers better in task performance and learning. Subjects' reaction to the received debriefing was also positive. As the subjects played two trials, the effect of practice was minimal. So, the overall improvement in the subject's decision making and learning can fairly be attributed to the debriefing. When it comes to the implementation of debriefing-based SDILEs, one has to balance the cost and benefits: with process-oriented debriefing, better performance can be expected, but it will require more resources (i.e., a trained facilitator's time) as compared with outcome-oriented debriefing.

References

Abdel-Hamid, T., Sengupta, K., & Swett, C. (1999). The impact of goals on software project management: An experimental investigation. *MIS Quarterly, 23*(4), 1–19.

Qudrat-Ullah, H. (2014). Yes we can: Improving performance in dynamic tasks. *Decision Support Systems, 61*, 23–33.

Chapter 5
Discussion and Conclusions

Abstract With limited resources, the delivery of affordable and reliable healthcare is increasingly becoming a difficult task for all nations across the globe. Decision makers in the healthcare domain in Canada are faced with the issue of seeking a balance between HIV/AIDS prevention and treatment spending. Therefore, we used Canadian case data in this study. Here, we present key limitations of this study, our major findings, implications of dynamic decision making research, and implications of improving practice in dynamic tasks in various domains including computer simulation-based education and training, aviation, healthcare, and policymaking. Based on the results reported in Chap. 4, here we will specifically discuss and argue why debriefing-based SDILE was effective in improving users' decision making and learning in dynamic tasks and why it did not help users to become "efficient decision makers." We will also talk about how the users perceived the utility of SIADH-ILE in improving their decision making and learning in the dynamic task.

Keywords Affordable and reliable healthcare · Dynamic decision making · Debriefing-based SDILE · Efficient decision makers · SIADH-ILE · Feedback loops · Time delays · Incentives · Decisional aid · HIV/AIDS prevention · Medical screening · HAART · Video arcade syndrome · Structure-behavior graphs

5.1 Introduction

Each one of us makes decisions, be they good or bad. However, organizations, where most of the decisions are strategic, can't afford poor decision making and sure enough, not for a long time. They spend substantial resources on employee education and training throughout their careers. The development and use of simulation-based training systems have always been in high demand. In the previous chapter, we presented the main results about the efficacy of such a solution – how did a SDILE, SIADH-ILE, impacts users' decision making and learning in a dynamic task. We found that debriefing, especially process-oriented debriefing, helped users do better on task performance, decision strategy, structural knowledge, and heuristic knowledge. Contrary to our expectation, in our laboratory experimental setting, subjects with process-oriented debriefing took a longer time in decision making (i.e., they were not efficient decision makers). Why did it happen? What caused some subjects' improved performance than the others? Why were the subjects with process-oriented debriefing not that efficient in their decision making? How do the users of SIADH-ILE perceive its utility in dynamic decision making? In this chapter, after describing the limitations of our study, we will address these questions to shed light on our results and conclusions.

5.2 Limitations

The main limitations of this study center on the use of the laboratory experimental approach and the participants. The experimental setting was computer simulation-based, SIADH-ILE, laboratory experimentation. Although there is a general concern of external validity associated with the experimental approach, computer simulation-based laboratory experimentation is advantageous in the study of complex dynamic tasks such as the management of HIV/AIDS task. In such simulated environments, much of the complexity in the tasks is the result of interactions among the flow of information, actions, and consequences during the task performance (Sterman 1989). In SDILEs, experimental simulations can capture some of the dynamic intertemporal aspects of tasks that are characteristic of natural settings but absent from preprogrammed stimuli in a controlled laboratory experiment (Abdel-Hamid et al. 1999). Also, a laboratory experimental approach is an established and authentic method of scientific inquiry within the fields of psychology and organizational behavior. A review of this literature reveals remarkable similarities between research findings obtained in the laboratory and field settings (Locke 1986; Qudrat-Ullah 2007). A priori, there is no reason to believe that the use of the experimental approach should be less successful to study dynamic decision making than to study judgment and choice in traditional cognitive psychology.

The use of students as surrogates raises the question of the generalizability of the results. However, student participants have been used in many dynamic decision making studies, including software project environments (Abdel-Hamid et al. 1999), production scheduling tasks and market strategy tasks (Sterman 2000), and fisheries management tasks (Moxnes 2004; Qudrat-Ullah 2014), on the basis of the general conclusion that the formal properties of tasks are much more important determinants of decision making in dynamic tasks than the participants' profiles (Ashton and Kramer 1980). Although the task of managing an HIV/AIDS task is somewhat different from all of these tasks, it is similar enough (e.g., it has feedback loops, time delays, nonlinear relationships) to assume that student participants are acceptable surrogates in this experimental investigation. Also, our participants had professional experience in healthcare ranging from 3 to 5 years which makes them practice-oriented adults rather than students per se. Thus, the use of adult learners in our experiments does not pose any threat to the results which are reported in Chap. 4.

In this experimental design, the lack of explicit incentives for all the participants deserves further explanation. Contrary to the common assumption that incentives lead to improved performance (Payne et al. 1993), findings in experimental economics suggest that this widely held belief is not always supported (Beattie & Looms, 1997). Similarly, empirical research on behavioral decision making indicates inclusive results on the impact of incentives. That is, incentives may increase, have no effect on, or at times actually decrease performance (Wright & Abdul-Ezz, 1988). Specifically, performance in complex, dynamic tasks is more a function of how interesting the tasks are rather than the monetary incentives (Paich and Sterman 1993; Sterman 2000; Qudrat-Ullah 2014). In the experiments in our study, interactions with the participants after debriefing revealed that they liked the task and did their best to accomplish it. When users are motivated in simulation-based decision making environments, they are expected to perform better.

5.3   Key Findings

This study demonstrates that debriefing is an effective decisional aid for dealing with complex, dynamic tasks. As the domain of our SDILE was healthcare, we report here some findings of dealing with limited resources for the prevention and treatment options for HIV/AID situation in Canada. The key findings are:

- Using SIAHD-ILE, we analyzed the long-term impacts of subjects' two important decision rules of an HIV/AIDS prevention and treatment strategy—(1) a reduction in infection ratio and (2) a reduction in prevalence ratio. Our analysis suggests that (i) a 22% reduction in infection ratio would reduce the number of infected women by 27% for those who are HIV positive due to heterosexual contact and by 17% for those who are AIDS positive due to heterosexual contact and (ii) a 22% reduction in prevalence ratio would reduce the number of infected women by 29% for those who are HIV positive due to heterosexual contact and by 18% for those who are AIDS positive due to heterosexual contact. These results provide strong support for a reexamination of the HIV prevention and treatment dichotomy, as has been strongly advocated by the UN Joint Programme on HIV/AIDS as part of a comprehensive combination prevention strategy (Montaner et al. 2010). Therefore, to decrease the number of HIV/AIDS infected women in Canada, Health Canada and its partners should commit resources to reduce simultaneously both infection rate and prevalence rate.
- Our results also support the assertion that modifying sexual attitude is one of the most important factors in decreasing the number of HIV/AIDS infected women. Investments in "education and awareness about HIV/AIDS" are well worth it— these would lead to reducing the infection rate among women substantially.

Similarly, the mandatory immigrant medical screening policy of Canada should continue (Krentza and Gill 2010). In fact, other countries can also explore the potential benefits of such policy measures.

The main assertion put forward in this book is about the efficacy of debriefing-based SDILEs in improving users' decision making and learning in dynamic tasks. Now, we present some key findings in this domain:

- The groups with any type of debriefing (i.e., process-oriented or outcome-oriented) showed significantly better performance on the dynamic task (i.e., managing HIV/AIDS task). It is worth noting that despite their better performance, none of these debriefed groups performed at the optimal level. In fact, they could not do better than the benchmark rule. These results are in line with several studies' findings in the dynamic decision making area (Plate 2010; Sterman 2000; Qudrat-Ullah 2014). In contrast to the famous "misperception of feedback hypothesis"—people perform poorly in dynamic tasks because they misperceive the effects of feedback (Sterman 1989, 2000)—our results, although modestly, show that with process-oriented debriefing subject performance in dynamic tasks can be improved.
- Training with process-oriented debriefing-based SDILE influenced subjects' decision strategies; subjects appear to adapt to systematic-variable decision strategies (i.e., fluctuations in their decisions are few and systematic) aimed at identifying decision patterns. On the other hand, subjects with no debriefing exhibited more erratic fluctuations in both their contact frequency decisions and HAART's effectiveness decisions. Subjects in the outcome-oriented debriefing group appear to adapt to systematic-variable decision strategies to a much lesser extent than did the process-oriented debriefing group.
- The effect of process-oriented debriefing on subjects' decision time (i.e., to make them efficient decision makers) was not as we expected. Subjects in the process-oriented debriefing group spent more time than any group. One explanation of this surprising result is that subjects in the process-oriented debriefing group adapted to the use of systematic-variable strategies to better understand the changes and patterns of the dynamic task. This kind of systematic, step-by-step, approach to decision making requires time. However, perhaps with more practice, the process-oriented debriefing group can become an efficient decision maker.
- Debriefing helped users improve their structural knowledge about the task system. Specifically, the provision of *process-oriented debriefing* appeared as the most successful decisional aid. Subjects with process-oriented debriefing developed better knowledge of the relationships between the variables of the task system because of their access to (i) information on the key task variables and on the nature of their relationships (e.g., delayed, nonlinear, and feedback orientation) and (ii) what Schön (1984) called "reflective conversation with the situation." Subjects with *outcome-oriented debriefing* appear to struggle in accounting for the changes in the task variables—an important condition to develop a better

understanding of dynamic tasks. Perhaps they focused on the expert solution provided to them and tried to mimic the performance.

- Debriefing also helped the participants develop heuristics. Again, the *process-oriented debriefing* had the most pronounced effect on the subject's heuristic knowledge. It appears that the availability of causal information through structure-behavior graphs made the difference. Subjects with outcome-oriented debriefing had to figure out the causal nature of the relationship between the key variables of the task system which often requires more practice than just the two trials they had. Process-oriented debriefing, on the other hand, gave the decision makers greater adaptability in recognizing task patterns and developing better heuristics.

- The effect of practice (i.e., playing two formal trials) was minimal. This is consistent with prior studies in dynamic decision making where it has been shown that even several rounds of doing the dynamic tasks will hardly improve users' performance. Performance improvement in dynamic tasks, instead, owes to subjects' understanding of the task structures (Moxnes 2004; Qudrat-Ullah 2014). Any support mechanism that helps the users to better understand the structures of the task system can be considered in the design of a learning environment.

- Human-human interactions during a debriefing session appear to be a necessary condition for the effective education and training of people in dynamic decision making. The ability of human facilitators/debriefers to address any unpredictable question(s) from the users and link them to appropriate structure-behavior casual links between the relevant variables of the task system makes them indispensable, at least for now before artificial intelligence (AI) reaches a much higher level.

- Perceived utility of SIADH-ILE was also assessed. Subjects rated the use of SIADH-ILE in improving and motivating them to do better in the dynamic task as very high. Business organizations as well as academic institutions can, therefore, embrace the use of SDILEs in the development and training programs of their employees and students.

- This research has advanced research on dynamic decision making by developing and using a comprehensive research model aimed at evaluating the effectiveness of debriefing on both the decision *processes* and their *outcomes*. This five-dimensional model (i.e., task performance, decision time, decision strategy, structural knowledge, and heuristic knowledge) can be utilized by researchers and practitioners to evaluate the effectiveness of any simulation-based decision support system.

- Although we evaluated the efficacy of SDILEs in improving users' decision making and learning in a dynamic task, SIADH-ILE, in the healthcare domain, the findings of this study are generic enough to be applicable to dynamic tasks in other domains. For instance, the training needed for people in aviation, medicine (e.g., emergency and surgical operations), cybersecurity, and organizational strategic decision making domains does require the education and development of expertise in dynamic tasks. The tested model of using debriefing-based SDILEs should equally benefit the communities of these domains.

5.4   Debriefing in Service of Dynamic Decision Making

This study is perhaps the first comprehensive study of structured debriefing-based SDILEs as decisional aids in dynamic tasks. In contrast to prior studies on dynamic decision making and effectiveness of ILEs, we were able to show significant improvements in decision processes and decision outcomes when decision makers in SDILEs were supported with process-oriented and outcome-oriented debriefing. Here we address some critical aspects of our results:

1. *In contrast to most of the prior studies in DDM, why did users of SIADH-ILE perform better in the dynamic task?*
2. *Contrary to our assertion, why did users of SIADH-ILE with process-oriented debriefing not become efficient decision makers?*
3. *What helped users of SIAD-ILE to overcome the video arcade syndrome?*

A sound discussion and analysis of these issues, we hope, will advance both the research and practice of dynamic decision making—an improvement of human performance in dynamic tasks.

5.4.1   Why Do People Perform Better in a Dynamic Task?

Why and how debriefing helped the users in SIADH-ILE? First, providing debriefing is useful because it improves subjects' task performance, structural knowledge, and heuristic knowledge and influences them to adapt to systematic-variable decision strategies. In concert with prior literature (Blazer et al. 1989; Plate 2010), our research shows the need to utilize these decisional aids—expert solution-oriented information and structure-behavior elucidating causal information about the task—in order to enhance decision makers' performance in dynamic tasks. In particular, the availability of process-oriented debriefing made the difference. Through the use of structure-behavior graphs and discussion about why performance was good or not good, learners were able to develop a better mental model of the task system and clarify any misconceptions about the structure of the dynamic task. In contrast to the learners with outcome-oriented debriefing who appear to mimic the performance of the expert solution that was shared with them during the debriefing session, process-oriented debriefing allowed users to develop pattern recognition skills. Without these additional decisional aids, the mere use of SDILEs did not help the subjects to improve on any measure of decision making and learning that was used in this study. It is interesting to note that subjects' performance in a simpler version of the dynamic task (with no delays and no feedback loops) in our practice trial, Trial 0, was quite satisfactory (i.e., all the groups had saved lives at a lower cost). However, when they faced the actual dynamic task in Trial 1, where they had the SDILE but no debriefing, their performance was significantly lower. Their subsequent performance improvement in Trial 2, thus, can reasonably be attributed to the debriefing.

Thus, it was not just the availability of mere additional information but the specific nature of information (e.g., clarity about the causality between various variables of the dynamic task system) that caused subjects' improved performance.

It is worth noting that despite the overwhelming evidence of the effects of debriefing on subjects' task performance (i.e., the total lives saved), none of them did (statistically) better than the benchmark rule. In fact, it would be too simplistic and naïve to think that subjects will become experts as a result of a couple of trials. On the other hand, according to Sternberg's theory on expertise development, people are experts on a continuum scale of expertise (Sternberg 1995): they are expert with varying degrees of expertise, some are at the beginner level, some are at a more mature level, and others are in between on the spectrum of expertise. This implies that we should look at broader measures of task performance in dynamic tasks. In fact, our results show that the subjects with debriefing-based training developed important skills including structural knowledge, heuristic knowledge, and the use of systematic pattern-seeking decision strategies.

5.4.2 Why Did Debriefing Not Help Decision Makers Become Efficient?

Contrary to our hypothesis, subjects with process-oriented debriefing did not improve on decision time. On the other hand, the very same subjects performed the best, as compared with the rest, on all of the other outcome measures: task performance, decision strategy, structural knowledge, and heuristic knowledge. This result clearly contrasts with the effort-outcome trade-off decision theory of Payne et al. (1990). There are at least two plausible explanations. First, this result appears to be consistent with the wisdom of general consulting services—inspiring clients how to fish (encouraging them to actively explore the relationships between task variables) rather than giving them the fish (providing the causal information about the central variables of the task system) (Qudrat-Ullah 2014). It appears that decision makers will benefit when providing them with the structure-behavior graph elucidating causal information that stimulates the exploration for decision heuristic and structural knowledge rather than by feeding them expert's rules. Second, in dynamic decision making, people's mental models are grossly simplified compared to reality. For instance, subjects don't appreciate the effect of time lags between an action (e.g., hiring people) and its consequence (e.g., availability of trained people). The development of such causal understanding among the variables of the task system requires active and systematic exploration and evaluation of one's decision rules. There are no shortcuts. Subjects with process-oriented debriefing appear to use the systematic-variable consistent decision strategies, which then resulted in their superior task performance. On the other hand, outcome-oriented debriefing group achieved some decision efficiency (i.e., less effort), but their performance was significantly inferior to those with process-oriented debriefing. On subjects' reaction to the debriefing they received, both the groups ranked the use of debriefing as very high. Thus, it is the availability and use of the type of debriefing (i.e., inducing causal understanding of the structure-behavior relationship among the task variables versus inducing heuristics based on the expert's decisions) rather than the cost-benefit trade-off decision strategies that resulted in improved performance. Perhaps more practice for those in the process-oriented debriefing group will make them efficient decision makers.

5.4.3 Overcoming the Video Arcade Syndrome

The effect of "practice" for the group with no debriefing on decision making and learning was insignificant. This is consistent with prior research findings (Kriz 2003; Spector 2000) that subjects in simulations, in the absence of a structured debriefing, often succumb to *video arcade syndrome*—they might win the game but without developing much understanding of the underlying task. Simply playing a simulation game might not be sufficient because users may view the game as a "black box" even if they implicitly understand how it works (Größler et al. 2016).

Instead, our results support the notion that additional resources in terms of a systematic and structured debriefing—embodying structure-behavior elucidating causal information about the task—should be used to develop a better understanding of dynamic tasks. More task transparency leads to subjects' improved performance.

Most prior studies on dynamic decision making (Paich and Sterman 1993; Diehl and Sterman 1995; Moxnes 2004) measured ILE's effectiveness on one or two criteria. Consequently, their conclusions are incomplete, if not biased. People's decision making and learning in simulation-based education and training can't be simply equated to "just doing the task." For instance, any learned tacit knowledge is difficult to articulate but it is there. By developing and using a model comprising of five evaluative criteria, this study has addressed the need to comprehensively evaluate the effectiveness of DSS, ILEs, and SDILEs. In this model, people's expertise in dynamic tasks ranging from a novice to a more mature expert level can be captured. Perhaps some of the prior studies can be re-evaluated using this five-dimensional model.

5.5 Improving Managerial Practice

Decision making and learning about dynamic tasks are not an easy endeavor. Organizations spend significant resources to educate and train their people for improved performance in dynamic tasks. Traditional semester-long formal education does improve performance in dynamic tasks (Sterman 2000). However, for today's fast-paced and task-intensive business organizations, affording this traditional time-consuming and prohibitively expensive formal education is a challenge. As a low-cost and efficient alternative, a 2.5-hour training session with a process-oriented debriefing-based SDILE in a workshop setting appears to be a viable solution to meet this managerial education and training need. Yes, we need to develop resources including a validated SDILE as well as trained debriefers/facilitators to conduct these workshops.

Our findings are beneficial for the computer simulation-based education and training industry which is a multibillion dollar industry (NTSA 2011). To accrue decision making skills—a *raison d'être* for education and training industry—practice with the simulation task alone won't be enough as our results suggest. Instead, deliberate and structured debriefing especially process-oriented debriefing is essential. With process-oriented debriefing at hand, subjects can better understand the task system and develop appropriate heuristics. It is the debriefing that helps the subjects better assimilate the new knowledge with their existing mental models and overcome the misperceptions about the task system they might have (Qudrat-Ullah 2007, 2010; Sterman 1994, 2000). Therefore, overwhelming confirmation of our hypothesis, "the group with process-oriented debriefing did the best on decision making and learning," should be valuable to the designers, developers, and users of simulation-based systems, who must decide which training method to implement to realize the gains desired from decision technologies (Qudrat-Ullah 2014). This finding is specifically useful for the aviation industry where the cost of simulator-based training is huge.

In the education sector, prior research has indicated that students were best able to develop a deeper understanding of the observed phenomena when they made connections between the micro- and macro-levels of the phenomena, while most school curricula deal with these phenomena in separate classes (Te'eni 1991). System dynamic-based ILEs, such as SIADH-ILE, have the potential to bridge this gap. Allowing users to see the global picture of the task system and understanding and linking structural elements to the relevant outcome or performance behaviors are the distinctive feature of SDILEs, which makes them stand out among other decisional aids. The education providers and administrators can encourage the development and use of structured and systematic debriefing-based SDILEs supported courses at appropriate levels in their program curricula.

In general, the theme of learning via simulations is a rich research area (Lakeh and Ghaffarzadegan 2015; Bell et al. 2008; Alessi and Kopainsky 2015; Qudrat-Ullah et al. 1997). This becomes even more pronounced a need when the cost associated with simulator-based training (e.g., pilot training with simulators) is prohibitively costly. Based on our positive outcomes with regard to people's learning (i.e., structural and heuristic knowledge) in dynamic tasks, researchers in the learning and instruction domain can benefit in several ways: (i) they can design new learning environments for integrated courses at upper levels, (ii) they can design and incorporate debriefing-based interventions in the existing simulation-based pedagogical tools, and (iii) they can evaluate various ILEs, SDILEs, and DSS for their efficacy in promoting users' decision making and learning in dynamic tasks.

Finally, policymakers in various domains such as education, health, military, transport, energy, where often stakes are really high, can make use of these SDILEs for the training of their people. Most of the policy decisions in these domains are strategic in nature. You make decisions today (i.e., spend resources) to see their impact later and sometimes often after several years or even decades. Before committing resources and making irreversible decisions, the hands-on practice with relevant SDILEs will help them gain insights in a safe and non-threatening environment and eventually make better policy decisions. In terms of having domain-specific SDILEs, companies can either develop them using in-house capabilities or just hire consultants to do the job. Regardless of the development logistics, in the development of an SDILE, the assumptions and perceptions of the decision makers are explicitly incorporated (or modeled) which facilitates the SDILE's ownership by the management—an essential aspect for the use and adaptation of SDILEs in any organization.

5.6 Summary of Discussion and Conclusions

We started this journey with the assertion that people's decision making in dynamic tasks can be improved. Based on our results as reported in Chap. 4, we have found significant support for this assertion: process-oriented debriefing-based training helps in people's decision making and learning in dynamic tasks. The availability of structure-behavior graphs appears to help users develop better "mental models" of the dynamic task system which in turn resulted in their improved performance. We employed a laboratory experimental approach to validate our research model. We developed and used a five-dimensional evaluative model for assessing the efficacy of SDILEs in improving subjects' dynamic decision making skills. Process-oriented debriefing helped users perform better in the task, develop systematic consistent pattern recognizing decision strategies, and use better heuristic and structural knowledge. We also explained how users were able to avoid video arcade syndrome. Researchers in the dynamic decision making field can benefit from the implications discussed here. Practitioners, especially in the simulation design and development sector, can also avail the utility of our tested five-dimensional evaluative model. Instead of restricting the measurement of subjects' performance to only task performance, they can value and asses people's performance on multiple dimensions including decision strategy, decision time, structural knowledge, and heuristic knowledge.

Overall, departing from the traditional focus of dynamic decision making research on "poor performance of people in dynamic tasks," we have showcased a modest intervention-based approach (i.e., the education and training with structured debriefing-based SDILEs) that helps people perform better in dynamic tasks in the range of skills. Depending upon the objectives and needs of your organization's training program, outcome-oriented and process-oriented debriefing-based SDILEs are available to improve the decision making of people in dynamic tasks. You the reader, after covering a good ground on the topics of this book, be the judge.

References

Abdel-Hamid, T., Sengupta, K., & Swett, C. (1999). The impact of goals on software project management: An experimental investigation. *MIS Quarterly, 23*(4), 1–19.

Alessi, A., & Kopainsky, B. (2015). System dynamics and simulation-gaming: Overview. *Simulation and Gaming, 48*(2–3), 223–229.

Ashton, R., & Kramer, S. (1980). Students as surrogates in behavioral accounting research: Some evidence. *Journal of Accounting Research, 18*(1), 15.

Beattie, J., & Looms, G. (1997). The impact of incentives upon risky choice experiments. *Journal of Risk and Uncertainty, 14*, 155–168.

Bell, B. S., Kanar, A. M., & Kozlowski, S. W. J. (2008). Current issues and future directions in simulation-based training in North America. *The International Journal of Human Resource Management, 19*(8), 1416–1434.

Blazer, W. K., Doherty, M. E., & O'Connor, R. (1989). Effects of cognitive feedback on performance. *Psychological Bulletin, 106*(3), 410–433.

Diehl, E., & Sterman, J. D. (1995). Effects of feedback complexity on dynamic decision making. *Organizational Behavior and Human Decision Processes, 62*(2), 198–215.

Größler, A., Rouwette, E., & Vennix, J. (2016). Non-conscious vs. deliberate dynamic decision-making—A pilot experiment. *Systems, 4*(13), 1–13. https://doi.org/10.3390/systems4010013.

Krentza, H., & Gill, M. (2010). The five-year impact of an evolving global epidemic, changing migration patterns, and policy changes in a regional Canadian HIV population. *Health Policy (Amsterdam), 90*(2–3), 296–302.

Kriz, W. C. (2003). Creating effective learning environments and learning organizations through gaming simulation design. *Simulation and Gaming, 34*(4), 495–511.

Lakeh, B., & Ghaffarzadegan, N. (2015). Does analytical thinking improve understanding of accumulation? *System Dynamics Review, 31*(1–2), 46–65.

Locke, E. A. (1986). *Generalizing from laboratory to field settings: Research findings from organizational behavior and human resource management*. Lexington, MA: Heath Lexington.

Montaner, J., Lima, V., Barrios, R., Yip, B., Wood, E., Kerr, T., & Kendall, P. (2010). Association of highly active antiretroviral therapy coverage, population viral load, and yearly new HIV diagnoses in British Columbia, Canada: A population-based study. www.thelancet.com, *376*, 532–539.

Moxnes, E. (2004). Misperceptions of basic dynamics: The case of renewable resource management. *System Dynamics Review, 20*, 139–162.

NTSA. (2011). President's notes. *Training Industry News, 23*(4), 2–2.

Paich, M., & Sterman, D. (1993). Boom, bust, and failures to learn in experimental markets. *Management Science, 39*(12), 1439–1458.

Payne, J. W., Johnson, E. J., Bettman, J. R., & Coupey, E. (1990). Understanding contingent choice: A computer simulation approach. *IEEE Transactions on Systems, Man, and Cybernetics, 20*, 296–309.

Payne, J. W., Bettman, J. R., & Johnson, E. J. (1993). *The adaptive decision maker*. New York: Cambridge University Press.

Plate, R. (2010). Assessing individuals' understanding of nonlinear causal structures in complex systems. *System Dynamics Review, 28*(1), 19–33.

Qudrat-Ullah, H. (2007). Debriefing can reduce misperceptions of feedback hypothesis: An empirical study. *Simulations and Gaming, 38*(3), 382–397.

Qudrat-Ullah, H. (2010). Perceptions of the effectiveness of system dynamics-based interactive learning environments: An empirical study. *Computers and Education, 55*, 1277–1286.

Qudrat-Ullah, H. (2014). Yes we can: Improving performance in dynamic tasks. *Decision Support Systems, 61*, 23–33.

Qudrat-Ullah, H., Saleh, M. M., & Bahaa, E. A. (1997). Fish Bank ILE: An interactive learning laboratory to improve understanding of 'The Tragedy of Commons'; a common behavior of complex dynamic systems. *Proceedings of 15th International System Dynamics Conference*, Istanbul, Turkey.

Schön, D. (1984). *The reflective practitioner*. New York: Basic Books.

Spector, J. M. (2000). System dynamics and interactive learning environments: Lessons learned and implications for the future. *Simulation and Gaming, 31*(4), 528–535.

Sterman, J. D. (1989). Modeling managerial behavior: Misperceptions of feedback in a dynamic decision making experiment. *Management Science, 35*, 321–339.

Sterman, J. D. (1994). Learning in and around complex systems. *System Dynamics Review, 10*, 291–323.

Sterman, J. D. (2000). *Business dynamics: Systems thinking and modeling for a complex world*. New York: McGraw-Hill.

Sternberg, R. J. (1995). Expertise in complex problem solving: A comparison of alternative conceptions. In P. Frensch & J. Funke (Eds.), *Complex problem solving: The European perspective* (pp. 3–25). Hillsdale, NJ: Lawrence Erlbaum Associates Publishers.

Te'eni, D. (1991). Feedback in DSS as a source of control: Experiments with the timing of feedback. *Decision Sciences, 22*(3), 644–655.

Wright, W. F., & Abdul-Ezz, M. E. (1988). Effects of extrinsic incentives on the quality of frequency assessments. *Organizational Behavior and Human Decision Processes, 41*, 143–152.

Chapter 6
Future Research Directions in Dynamic Decision Making

Abstract The cumulative knowledge of what works and what doesn't work in the education and training of people in dynamic tasks is of immense importance. Simulation-based education and training is a multibillion industry, and stakes are high (e.g., we need to train future surgeons, aviators, and the policymakers). The ILEs in general and SDILEs, in particular, have long been used and are considered a viable solution to this ever-increasing need of people's education and training in dynamic decision making. In this book, we have modestly contributed to advance the science of dynamic decision making and also to the sense of evaluation of decisional aids. However, we do admit that there are more questions to be investigated than what we have addressed here.

Keywords Cumulative knowledge · Evaluation science · Dynamic decision making · Cost-benefit analysis · Effort-reward hypothesis · Debriefing models · Prevention and treatment options · Management of HIV/AIDS situation · Transfer learning: Health professionals · CO_2 accumulation

6.1 Introduction

It is not uncommon for a research project, after its completion, to have more new research topics and questions that are answered in that project. Our book project is not an exception. To begin with, DDM and learning in and with SDILEs are very research areas. When system dynamic models or SDILEs are applied to various domains to better understand the dynamic tasks in that domain, often counterintuitive insights are revealed with some interesting implications for research and practice.

Although the concept of debriefing is not new at all, the systematic incorporation of debriefing in the design of SDILEs is relatively new area itself. For instance, we incorporated and tested the validity of only two types of debriefing in a group setting, other forms and modalities are for future research to explore. Similarly, we applied our SDILE, SIADH-ILE, to a Canadian case; sure only more replicative studies can increase the appeal and validity of our findings. In the following section on "Potential Research Avenue," we highlight some specific research problems that researchers can immediately avail and being their research projects.

6.2   Potential Research Avenues

Here, we mention some interesting future research avenues:

- Although we have found strong support for the use of process-oriented debriefing in enhancing people's task performance, decision strategies, heuristics, and structural knowledge, any cost-benefit analysis was not performed. Instead of using automated, SDILE help system-based outcome-oriented debriefing, process-oriented debriefing requires the expertise of a human debriefer or facilitator. Both time and money are needed. Depending on the objectives of the training with SDILEs, more or less structural understanding inducing sessions with outcome-oriented debriefing or process-oriented debriefing can be designed.
- Contrary to the effort-reward hypothesis, people spend effort proportional to the expected reward, and the users in our SDILE with process-oriented debriefing spent more time to improve their structural learning about the task system and were not efficient. Perhaps more practice for this group (PD) would lead to make them efficient decision makers. Note that the PD group was more accurate in their decisions. So, it would be interesting to investigate: where is the balance in accuracy-efficiency continuum for debriefing-based training with SDILEs and what would be the associated costs?
- We applied Lederman's model (1992) in this study. An analysis and evaluation of other debriefing models using our tested five-dimensional evaluative model would be beneficial to compare with other educational methods, techniques, and interventions.
- The level of difficulty of dynamic task matters. Although our dynamic task, management of HIV/AIDS situation with prevention and treatment options, is semantically similar to other dynamic tasks such as FishBankILE (Qudrat-Ullah et al. 1997), C-Learn (Sterman, 2000; Sterman et al. 2014), Beer Game, and Boom and Bust, the efficacy of debriefing can be solidly generalized by using the laboratory-experiment as ours but with different SDILE (e.g., with more or less feedback loops that are in SIADH-ILE). Replicative studies of this nature need to further authenticate the efficacy of debriefing in improving people's decision making and learning in dynamic tasks.
- Prior studies on dynamic decision making seem to suffer in many respects. Often only a single measure of task performance is used, they rarely focus on decision making process, and the employed SDILEs are often developed without any

consideration of how people actually learn? Each of these individual research efforts and the field of dynamic decision making and learning in and with SDILEs as a whole should benefit from an integrated approach. Our demonstrated laboratory experimental approach effectively represents this integrated approach to learning and decision making in dynamic tasks. By empirically validating our model with more studies in different contexts, we could then compare results, build a cumulative knowledge base of effective decisional aids, and study the trade-offs among various kinds of decisional aids to learning and decision making in dynamic tasks. We are already engaged in such replicative research by using another SDILE, FishBankILE (Qudrat-Ullah et al. 1997; Qudrat-Ullah 2014).

- A critical dimension of any education and training activity, with or without simulation, is transfer learning: subjects should be able to learn and glean insights from simulation-based training and then perform actual tasks in the field. Although a difficult endeavor as it is, a follow-up field study, say with health professionals on the job, is worth pursuing.
- We evaluated the efficacy of collective debriefing only. Future studies can test the effectiveness of individual debriefing in enhancing people's decision making and learning in dynamic tasks.

6.3 Finally

Here I want to share with you my personal reflections on this work. Yes, I am passionate about improving people decision making both in personal and professional contexts. Being trained in system dynamics, I fully appreciate its utility in imparting "structure drive performance," a solid and proved mechanism to improve people's skills in managing dynamic tasks. However, in the corporate world and policymaker's worlds, it is hard if not impossible to send their people for the formal semester or more long study in system dynamics. Instead, workshop setting-based training sessions suite them. My curiosity to design and test such a workshop-based intervention to train people in dynamic decision making has culminated to this book where the efficacy of both outcome-oriented debriefing and process-oriented debriefing-based SDILE is empirically validated. In fact, I did test the effect of debriefing and published the results in the journal: Simulation and Gaming in 2007 (Qudrat-Ullah 2007). Since then the journey continues.

Most recently, we studied the effects of debriefing in the contest of a dynamic task in the climate change domain. Here too, it was widely reported in the literature the people misperceive the dynamics of CO_2 accumulation. We conducted a laboratory-based experiment using the dynamic task, C-Learn (Sterman 2000; Sterman et al. 2014). We found very encouraging results: when people were trained with the active briefing and debriefing-based SDILE, their performance and understanding about CO_2 accumulation improved significantly (Qudrat-Ullah and Kayal 2018). With these examples at hand, I hope, you the reader will fully avail and

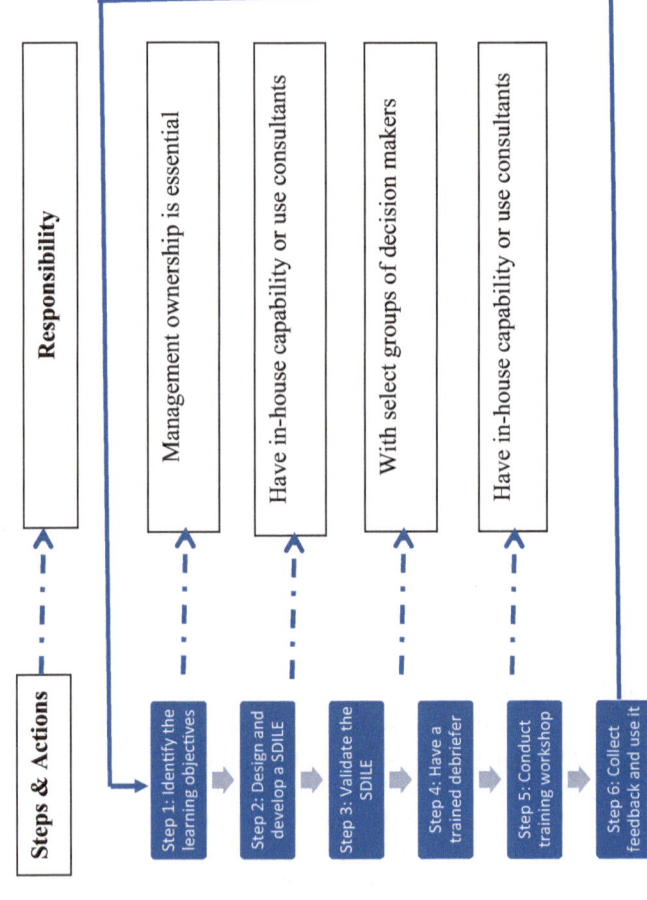

Fig. 6.1 A model to implement SDILE-based training in organizations

embrace the education and training of your management team with SDILEs in your organizations. If you any help in this regard, you feel free to contact me at hassnq@yorku.ca.

Here I share with you a step-by-step approach to implement SDILE-based education and training in your organizations (see Fig. 6.1).

Please note that these steps can be interactive and iterative in nature. Each step is unique and its related cost is different. You can focus more or less on any of these steps as it suite to your own situation.

6.4 Summary of Future Research Directions for DDM and Healthcare Domains

Improving human performance in dynamic tasks was the key objective of this book project. By now you, who has covered a good amount of material ranging from definitions of concepts to experimental investigation and validation of advanced prepositions about the utility of the debriefing-based SDILE, should be able to give the verdict on to what extent this book has met the stated objective. Nevertheless, research on any topic is an ongoing activity, so is the case for the research reported in this book. We have identified some specific research topics for researchers both in DDM area and healthcare (HIV/AIDs) domain. We are specifically hopeful for the future research pertaining to the various shapes, forms, models, and modalities of debriefing, learning theories and principles, and the design theory and principles—all aimed at improving the efficacy of SDILEs in promoting people' decision making in dynamic tasks.

References

Qudrat-Ullah, H. (2007). Debriefing can reduce misperceptions of feedback hypothesis: An empirical study. *Simulations and Gaming, 38*(3), 382–397.

Qudrat-Ullah, H. (2014). Yes we can: Improving performance in dynamic tasks. *Decision Support Systems, 61*, 23–33.

Qudrat-Ullah, H., & Kayal, A. (2018). How to improve learners' (mis) understanding of CO2 accumulations through the use of human-facilitated interactive learning environments? *Journal of Cleaner Production, 184*, 188–197.

Qudrat-Ullah, H., Saleh, M. M., & Bahaa, E. A. (1997). Fish Bank ILE: An interactive learning laboratory to improve understanding of 'The Tragedy of Commons'; a common behavior of complex dynamic systems. *Proceedings of 15th International System Dynamics Conference*, Istanbul, Turkey.

Sterman, J. D. (2000). *Business dynamics: Systems thinking and modeling for a complex world*. New York: McGraw-Hill.

Sterman, J., Franck, T., Fiddaman, T., Jones, A., McCauley, S., Rice, P., Sawin, E., Siegel, L., & Rooney-Varga, J. (2014). World climate: A role-play simulation of climate negotiations. *Simulation & Gaming*, 1–35.

A Thematic Bibliography

Here we present some topical and highly useful research pertaining to improving decision making in dynamic tasks. We have identified nine themes, (i) Laboratory Experimental Approach, (ii) Debriefing and Human Facilitation, (iii) Decision Making and Learning in Dynamic Tasks, (iv), Simulation-Based Training and Tools, (v) Dynamic Decision Making, (vi) Learning Theories and Principles, (vii) System Dynamics and ILEs, (viii) Designs Principle for Learning in and with Simulations, and (ix) Healthcare and Its Dynamics, so the readers could avail this literature according to their interest in a specific theme. We hope both the researchers and practitioners can benefit from this thematic bibliography.

1. Laboratory Experimental Approach

Abdel-Hamid, T., Sengupta, K., & Swett, C. (1999). The impact of goals on software project management: An experimental investigation. *MIS Quarterly, 23*(4), 1–19.

Blazer, W. K., Doherty, M. E., & O'Connor, R. (1989). Effects of cognitive feedback on performance. *Psychological Bulletin, 106*(3), 410–433.

Diehl, E., & Sterman, J. D. (1995). Effects of feedback complexity on dynamic decision making. *Organizational Behavior and Human Decision Processes, 62*(2), 198–215.

Fischer, H., & Gonzalez, C. (2015). Making sense of dynamic systems: How our understanding of stocks and flows depends on a global perspective. *Cognitive Science, 40*, 496–512.

Ford, D. N., & Mccormack, D. E. M. (2000). Effects of time scale focus on system understanding in decision support systems. *Simulation and Gaming, 31*(3), 309–330.

Gonzalez, M., Machuca, J., & Castillo, J. (2000). A transparent-box multifunctional simulator of competing companies. *Simulation and Gaming, 31*(2), 240–256.

Größler, A., Rouwette, E., & Vennix, J. (2016). Non-conscious vs. deliberate dynamic decision-making—A pilot experiment. *Systems, 4*(13), 1–13. https://doi.org/10.3390/systems4010013.

Howie, E., Sy, S., Ford, L., & Vicente, K. J. (2000). Human-computer interface design can reduce misperceptions of feedback. *System Dynamics Review, 16*(3), 151–171.

Kleinmuntz, D. (1985). Cognitive heuristics and feedback in a dynamic decision environment. *Management Science, 31*, 680–701.

Kottermann, E., Davis, D., & Remus, E. (1995). Computer-assisted decision making: Performance, beliefs, and illusion of control. *Organizational Behavior and Human Decision Processes, 57*, 26–37.

© The Author(s), under exclusive license to Springer Nature Switzerland AG 2020
H. Qudrat-Ullah, *Improving Human Performance in Dynamic Tasks*,
SpringerBriefs in Complexity, https://doi.org/10.1007/978-3-030-28166-3

Lakeh, B., & Ghaffarzadegan, N. (2015). Does analytical thinking improve understanding of accumulation? *System Dynamics Review, 31*(1–2), 46–65.

Locke, E. A. (1986). *Generalizing from laboratory to field settings: Research findings from organizational behavior and human resource management*. Lexington, MA: Heath Lexington.

Moxnes, E. (2004). Misperceptions of basic dynamics: The case of renewable resource management. *System Dynamics Review, 20*, 139–162.

Paich, M., & Sterman, D. (1993, December). Boom, bust, and failures to learn in experimental markets. *Management Science, 39*(12), 1439–1458.

Qudrat-Ullah, H. (2007). Debriefing can reduce misperceptions of feedback hypothesis: An empirical study. *Simulations and Gaming, 38*(3), 382–397.

Qudrat-Ullah, H. (2010). Perceptions of the effectiveness of system dynamics-based interactive learning environments: An empirical study. *Computers and Education, 55*, 1277–1286.

Qudrat-Ullah, H. (2014). Yes we can: Improving performance in dynamic tasks. *Decision Support Systems, 61*, 23–33.

Qudrat-Ullah, H., & Kayal, A. (2018). How to improve learners' (mis) understanding of CO2 accumulations through the use of human-facilitated interactive learning environments? *Journal of Cleaner Production, 184*, 188–197.

Sengupta, K., & Abdel-Hamid, T. (1993). Alternative concepts of feedback in dynamic decision environments: An experimental investigation. *Management Science, 39*, 411–428.

Sterman, J. D. (1989). Modeling managerial behavior: Misperceptions of feedback in a dynamic decision making experiment. *Management Science, 35*, 321–339.

Sterman, J. D., & Dogan, G. (2015). I'm not hoarding, I'm just stocking up before the hoarders get there. Behavioral causes of phantom ordering in supply chains. *Journal of Operations Management, 39-40*, 6–22.

Sterman, J. D., & Sweeney, B. (2007). Understanding public complacency about climate change: Adults' mental models of climate change violate conservation of matter. *Climatic Change, 80*(3–4), 213–238.

Sterman, J., Franck, T., Fiddaman, T., Jones, A., McCauley, S., Rice, P., Sawin, E., Siegel, L., & Rooney-Varga, J. (2014). World climate: A role-play simulation of climate negotiations. *Simulation & Gaming, 46*(3): 1–35.

Te'eni, D. (1991). Feedback in DSS as a source of control: Experiments with the timing of feedback. *Decision Sciences, 22*(3), 644–655.

Yi, S., & Davis, F. (2001). Improving computer training effectiveness for decision technologies: Behavior modeling and retention enhancement. *Decision Sciences, 32*(3), 521–544.

Tamara, E. F., Alec, M. B., & Judith, D. S. (2013). The use of technology by nonformal environmental educators. *The Journal of Environmental Education, 44*(1), 16–37.

2. Debriefing and Human Facilitation

Allen, J. A., Baran, B. E., & Scott, C. W. (2010). After-action reviews: Avenue for the promotion of safety climate. *Accident Analysis and Prevention, 42*, 750–757.

Dismukes, R. K., & Smith, G. M. (Eds.). (2000). *Facilitation in aviation training and operations*. Aldershot, UK: Ashgate.

Dreifuerst, K. T. (2009). The essentials of debriefing in simulation learning: A concept analysis. *Nursing Education Perspectives, 30*(2), 109–114.

Ellis, S., & Davidi, I. (2005). After-event reviews: Drawing lessons from successful and failed experience. *Journal of Applied Psychology, 90*, 857–871.

Fanning, M., & Gaba, M. (2007). The role of debriefing in simulation-based learning. *Simulation in Healthcare, 2*(2), 115–125.

Kim, J., & Pavlov, O. (2017). Game-based structural debriefing: A design tool for systems thinking curriculum. *SSRN Electronic Journal*. https://doi.org/10.2139/ssrn.3218674.

Lederman, L. C. (1992). Debriefing: Towards a systematic assessment of theory and practice. *Simulation and Gaming, 23*(2), 145–160.

Pavlov, O., Saeed, K., & Robinson, L. (2015, February). Improving instructional simulation with structural debriefing. *Simulation and Gaming, 46*, 383–403.

Qudrat-Ullah, H. (2007). Debriefing can reduce misperceptions of feedback hypothesis: An empirical study. *Simulations and Gaming, 38*(3), 382–397.

Qudrat-Ullah, H., & Kayal, A. (2018). How to improve learners' (mis) understanding of CO_2 accumulations through the use of human-facilitated interactive learning environments? *Journal of Cleaner Production, 184*, 188–197.

Qudrat-Ullah, H., & Kayal, A. (2018). How to improve learners' (mis) understanding of CO_2 accumulations through the use of human-facilitated interactive learning environments? *Journal of Cleaner Production, 184*, 188–197.

Rogers, E. W., & Milam, J. (2004). Pausing for learning: Applying the after action review process at the NASA Goddard Space Flight Center. In *Proceedings of the IEEEAC* (pp. 4383–4388). Piscataway, NJ: Institute of Electrical and Electronics Engineers.

Ron, N., Lipshitz, R., & Popper, M. (2002). How organizations learn: Post-flight reviews in a F-16 fighter squadron. *Organization Studies, 27*, 1069–1089.

Tannenbaum, S. I., & Cerasoli, C. P. (2013). Do team and individual debrief enhance performance? A meta-analysis. *Human Factors, 55*(1), 231–245.

3. Decision Making and Learning in Dynamic Tasks

Bakken, B. E. (1993). *Learning and transfer of understanding in dynamic decision environments.* Unpublished doctoral dissertation, MIT, Boston.

Barnett, S., & Ceci, S. (2002). When and where do we apply what we learn? A taxonomy for far transfer. *Psychological Bulletin, 128*(4), 612–637.

Berry, D. C. (1991). The role of action in implicit learning. *Quarterly Journal of Experimental Psychology, 43A*, 881–906.

Cohen, I. (2008). Improving time-critical decision making in life-threatening situations: Observations and insights. *Decision Analysis, 5*(2), 100–110.

Fischer, H., & Gonzalez, C. (2015). Making sense of dynamic systems: How our understanding of stocks and flows depends on a global perspective. *Cognitive Science, 40*, 496–512.

Gagné, R. M. (1985). *The conditions of learning and theory of instruction*. New York: Holt, Rinehart, and Winston.

Gagné, R. M., Briggs, L. J., & Wager, W. W. (1992). *Principles of instructional design* (4th ed.). Fort Worth, TX: Harcourt Brace Jovanovich College Publishers.

Garris, R., Ahlers, R., & Driskell, J. E. (2002). Games, motivation, and learning: A research and practice model. *Simulation and Gaming, 33*(4), 441–467.

Issacs, W., & Senge, P. (1994). Overcoming limits to learning in computer-based learning environments. In J. Morecroft & J. Sterman (Eds.), *Modeling for learning organizations* (pp. 267–287). Portland, OR: Productivity Press.

Lane, D. C. (1995). On a resurgence of management simulations and games. *Journal of the Operational Research Society, 46*, 604–625.

Mayer, W., Dale, K., Fraccastoro, K., & Moss, G. (2011). Improving transfer of learning: Relationship to methods of using business simulation. *Simulation & Gaming, 42*(1), 64–84.

Qudrat-Ullah, H. (2007). Debriefing can reduce misperceptions of feedback hypothesis: An empirical study. *Simulations and Gaming, 38*(3), 382–397.

Qudrat-Ullah, H., & Kayal, A. (2018). How to improve learners' (mis) understanding of CO2 accumulations through the use of human-facilitated interactive learning environments? *Journal of Cleaner Production, 184*, 188–197.

Rogers, E. W., & Milam, J. (2004). Pausing for learning: Applying the after action review process at the NASA Goddard Space Flight Center. In *Proceedings of the IEEEAC* (pp. 4383–4388). Piscataway, NJ: Institute of Electrical and Electronics Engineers.

Ron, N., Lipshitz, R., & Popper, M. (2002). How organizations learn: Post-flight reviews in a F-16 fighter squadron. *Organization Studies, 27*, 1069–1089.

Sengupta, K., & Abdel-Hamid, T. (1993). Alternative concepts of feedback in dynamic decision environments: An experimental investigation. *Management Science, 39*, 411–428.

Sterman, J. D. (1989). Modeling managerial behavior: Misperceptions of feedback in a dynamic decision making experiment. *Management Science, 35*, 321–339.

Sterman, J. D., & Dogan, G. (2015). I'm not hoarding, I'm just stocking up before the hoarders get there. Behavioral causes of phantom ordering in supply chains. *Journal of Operations Management, 39–40*, 6–22.

Sterman, J. D., & Sweeney, B. (2007). Understanding public complacency about climate change: Adults' mental models of climate change violate conservation of matter. *Climatic Change, 80*(3–4), 213–238.

Sterman, J., Franck, T., Fiddaman, T., Jones, A., McCauley, S., Rice, P., Sawin, E., Siegel, L., & Rooney-Varga, J. (2014). World climate: A role-play simulation of climate negotiations. *Simulation & Gaming*, 1–35.

4. Simulation-Based Training and Tools

Bell, B. S., Kanar, A. M., & Kozlowski, S. W. J. (2008). Current issues and future directions in simulation-based training in North America. *The International Journal of Human Resource Management, 19*(8), 1416–1434.

Faria, A. J. (1998). Business simulation games: Current usage levels – An update. *Simulation and Gaming, 29*, 295–308.

Fischer, H., & Gonzalez, C. (2015). Making sense of dynamic systems: How our understanding of stocks and flows depends on a global perspective. *Cognitive Science, 40*, 496–512.

Gaba, D. M., Howard, S. K., Fish, K. J., Smith, B. E., & Sowb, Y. A. (2001). Simulation-based training in anesthesia crisis resource management (ACRM): A decade of experience. *Simulation and Gaming, 32*, 175–193.

Lane, D. C. (1995). On a resurgence of management simulations and games. *Journal of the Operational Research Society, 46*, 604–625.

Lane, M., & Tang, Z. (2000). Effectiveness of simulation training on transfer of statistical concepts. *Journal of Educational Computing Research, 22*(4), 383–396.

Maier, F. H., & Größler, A. (2000). What are we talking about? – A taxonomy of computer simulations to support learning. *System Dynamics Review, 16*(2), 135–148.

Mayer, W., Dale, K., Fraccastoro, K., & Moss, G. (2011). Improving transfer of learning: Relationship to methods of using business simulation. *Simulation & Gaming, 42*(1), 64–84.

NTSA. (2011). President's notes. *Training Industry News, 23*(4), 2–2.

Spector, J. M. (2000). System dynamics and interactive learning environments: Lessons learned and implications for the future. *Simulation and Gaming, 31*(4), 528–535.

Qudrat-Ullah, H. (2008). Behavior validity of a simulation model for sustainable development. *International Journal of Management and Decision Making, 9*(2), 129–139.

Qudrat-Ullah, H., & BaekSeo, S. (2010). How to do structural validity of a system dynamics type simulation model: The case of an energy policy model. *Energy Policy, 38*(5), 2216–2224.

Sfard, A. (1998). On two metaphors for learning and dangers of choosing just one. *Educational Research, 27*(2), 4–12.

Sing, D. T. (1998). Incorporating cognitive aids into decision support systems: The case of the strategy execution process. *Decision Support Systems, 24*, 145–163.

Söllner, A., Brödery, A., & Hilbig, E. (2013). Deliberation versus automaticity in decision making: Which presentation format features facilitate automatic decision making? *Judgment and Decision making, 8*(3), 278–298.

Söllner, A., Brödery, A., & Hilbig, E. (2013). Deliberation versus automaticity in decision making: Which presentation format features facilitate automatic decision making? *Judgment and Decision making, 8*(3), 278–298.

Sterman, J. D. (1992). Teaching takes off – Flight simulators for management education, *OR/MS Today*, pp. 40–44.

Sterman, J., Franck, T., Fiddaman, T., Jones, A., McCauley, S., Rice, P., Sawin, E., Siegel, L., & Rooney-Varga, J. (2014). World climate: A role-play simulation of climate negotiations. *Simulation & Gaming*, 1–35.

Sternberg, R. J. (1995). Expertise in complex problem solving: A comparison of alternative conceptions. In P. Frensch & J. Funke (Eds.), *Complex problem solving: The European perspective* (pp. 3–25). NJ: Lawrence Erlbaum Associates Publishers.

Sternberg, R. J., & Horvath, J. A. (1995). A prototype view of expert teaching. *Educational Researcher, 24*(6), 9–17.

Sweller, J. (1988). Cognitive load during problem solving: Effects on learning. *Cognitive Science., 12*(2), 257–285.

USDOA. (1993). US Department of the Army, Military Operations: U.S. Army Operations Concept for Combat Identification, TRADOC Pam 525–58 (Fort Monroe, Va.: Training and Doctrine Command, 31 August 1993):1.

5. Dynamic Decision Making

Abdel-Hamid, T., Sengupta, K., & Swett, C. (1999). The impact of goals on software project management: An experimental investigation. *MIS Quarterly, 23*(4), 1–19.

Brehmer, B. (1990). Strategies in real-time dynamic decision making. In R. M. Hogarth (Ed.), *Insights in decision making* (pp. 262–279). Chicago: University of Chicago Press.

Diehl, E., & Sterman, J. D. (1995). Effects of feedback complexity on dynamic decision making. *Organizational Behavior and Human Decision Processes, 62*(2), 198–215.

Edwards, W. (1962). Dynamic decision theory and probabilistic information processing. *Human Factors, 4*, 59–73.

Fischer, H., & Gonzalez, C. (2015). Making sense of dynamic systems: How our understanding of stocks and flows depends on a global perspective. *Cognitive Science, 40*, 496–512.

Ford, D. N., & Mccormack, D. E. M. (2000). Effects of time scale focus on system understanding in decision support systems. *Simulation and Gaming, 31*(3), 309–330.

Gonzalez, M., Machuca, J., & Castillo, J. (2000). A transparent-box multifunctional simulator of competing companies. *Simulation and Gaming, 31*(2), 240–256.

Größler, A., Rouwette, E., & Vennix, J. (2016). Non-conscious vs. deliberate dynamic decision-making—A pilot experiment. *Systems, 4*(13), 1–13. https://doi.org/10.3390/systems4010013.

Hogarth, R. M. (1981). Beyond discrete biases: Functional and dysfunctional aspects of judgmental heuristics. *Psychological Bulletin, 9*(2), 197–217.

Howie, E., Sy, S., Ford, L., & Vicente, K. J. (2000). Human-computer interface design can reduce misperceptions of feedback. *System Dynamics Review, 16*(3), 151–171.

Keeney, R. L., & Raiffa, H. (1976). *Decisions with multiple objectives*. New York: Wiley.

Kleinmuntz, D. (1985). Cognitive heuristics and feedback in a dynamic decision environment. *Management Science, 31*, 680–701.

Lakeh, B., & Ghaffarzadegan, N. (2015). Does analytical thinking improve understanding of accumulation? *System Dynamics Review, 31*(1–2), 46–65.

Moxnes, E. (2004). Misperceptions of basic dynamics: The case of renewable resource management. *System Dynamics Review, 20*, 139–162.

Paich, M., & Sterman, D. (1993, December). Boom, bust, and failures to learn in experimental markets. *Management Science, 39*(12), 1439–1458.

Payne, J. W., Bettman, J. R., & Johnson, E. J. (1993). *The adaptive decision maker.* New York: Cambridge University Press.

Payne, J. W., Johnson, E. J., Bettman, J. R., & Coupey, E. (1990). Understanding contingent choice: A computer simulation approach. *IEEE Transactions on Systems, Man, and Cybernetics, 20*, 296–309.

Qudrat-Ullah, H. (2007). Debriefing can reduce misperceptions of feedback hypothesis: An empirical study. *Simulations and Gaming, 38*(3), 382–397.

Qudrat-Ullah, H. (2010). Perceptions of the effectiveness of system dynamics-based interactive learning environments: An empirical study. *Computers and Education, 55*, 1277–1286.

Qudrat-Ullah, H. (2014). Yes we can: Improving performance in dynamic tasks. *Decision Support Systems, 61*, 23–33.

Qudrat-Ullah, H., & Kayal, A. (2018). How to improve learners' (mis) understanding of CO2 accumulations through the use of human-facilitated interactive learning environments? *Journal of Cleaner Production, 184*, 188–197.

Rogers, E. W., & Milam, J. (2004). Pausing for learning: Applying the after action review process at the NASA Goddard Space Flight Center. In *Proceedings of the IEEEAC* (pp. 4383–4388). Piscataway, NJ: Institute of Electrical and Electronics Engineers.

Ron, N., Lipshitz, R., & Popper, M. (2002). How organizations learn: Post-flight reviews in a F-16 fighter squadron. *Organization Studies, 27*, 1069–1089.

Sengupta, K., & Abdel-Hamid, T. (1993). Alternative concepts of feedback in dynamic decision environments: An experimental investigation. *Management Science, 39*, 411–428.

Sterman, J. D. (1989). Modeling managerial behavior: Misperceptions of feedback in a dynamic decision making experiment. *Management Science, 35*, 321–339.

Sterman, J. D., & Dogan, G. (2015). I'm not hoarding, I'm just stocking up before the hoarders get there. Behavioral causes of phantom ordering in supply chains. *Journal of Operations Management, 39–40*, 6–22.

Sterman, J. D., & Sweeney, B. (2007). Understanding public complacency about climate change: Adults' mental models of climate change violate conservation of matter. *Climatic Change, 80*(3–4), 213–238.

Sterman, J., Franck, T., Fiddaman, T., Jones, A., McCauley, S., Rice, P., Sawin, E., Siegel, L., & Rooney-Varga, J. (2014). World climate: A role-play simulation of climate negotiations. *Simulation & Gaming*, 1–35.

6. Learning Theories and Principles

Briggs, P. (1990). Do they know what they are doing? An evaluation of word-processor user's implicit and explicit task-relevant knowledge, and its role in self-directed learning. *International Journal of Man-Machine Studies, 32*, 385–298.

Rouwette, A., Großler, A., & Vennix, M. (2004). Exploring influencing factors on rationality: A literature review of dynamic decision-making studies in system dynamics. *Systems Research and Behavioral Science, 21*, 351–370.

Collins, A. (1991). Cognitive apprenticeship and instructional technology. In L. Idol & B. F. Jones (Eds.), *Educational values and cognitive instruction: Implications for reform* (pp. 121–138). Hillsdale, NJ: Lawrence Erlbaum Associates.

Conant, R., & Ashby, W. (1970). Every good regulator of a system must be a model of the system. *International Journal of System Science, 1*, 89–97.

Cox, R. J. (1992). Exploratory learning from computer-based systems. In S. Dijkstra, H. P. M. Krammer, & J. J. G. van Merrienboer (Eds.), *Instructional models in computer-based learning environments* (pp. 405–419). Berlin, Heidelberg: Springer-Verlag.

Davidsen, P. I., & Spector, J. M. (1997). Cognitive complexity in system dynamics based learning environments. *International System Dynamics Conference*, Istanbul, Turkey: Bogazici University Printing Office, pp. 757–760.

Glöckner, A., & Betsch, T. (2008). Multiple-reason decision making based on the automatic processing. *Journal of Experimental Psychology: Learning, Memory, and Cognition, 34*, 1055–1075.

Kwakkel, J. H., & Pruyyt, E. (2013). Explanatory modeling and analysis and approach for model-based foresight under deep uncertainty. *Technological Forecasting and Social Change, 80*(3), 419–431.

Schön, D. (1984). *The reflective practitioner*. New York: Basic Books.

Sfard, A. (1998). On two metaphors for learning and dangers of choosing just one. *Educational Research, 27*(2), 4–12.

Sing, D. T. (1998). Incorporating cognitive aids into decision support systems: The case of the strategy execution process. *Decision Support Systems, 24*, 145–163.

Söllner, A., Brödery, A., & Hilbig, E. (2013). Deliberation versus automaticity in decision making: Which presentation format features facilitate automatic decision making? *Judgment and Decision making, 8*(3), 278–298.

Söllner, A., Brödery, A., & Hilbig, E. (2013). Deliberation versus automaticity in decision making: Which presentation format features facilitate automatic decision making? *Judgment and Decision making, 8*(3), 278–298.

Spector, J. M. (2000). System dynamics and interactive learning environments: Lessons learned and implications for the future. *Simulation and Gaming, 31*(4), 528–535.

Sterman, J. D. (1994). Learning in and around complex systems. *System Dynamics Review, 10*.

Sterman, J. D. (1992). Teaching takes off – Flight simulators for management education, *OR/MS Today*, pp. 40–44.

Sternberg, R. J. (1995). Expertise in complex problem solving: A comparison of alternative conceptions. In P. Frensch & J. Funke (Eds.), *Complex problem solving: The European perspective* (pp. 3–25). NJ: Lawrence Erlbaum Associates Publishers.

Sternberg, R. J., & Horvath, J. A. (1995). A prototype view of expert teaching. *Educational Researcher, 24*(6), 9–17.

Sweller, J. (1988). Cognitive load during problem solving: Effects on learning. *Cognitive Science, 12*(2), 257–285.

Mayer, W., Dale, K., Fraccastoro, K., & Moss, G. (2011). Improving transfer of learning: Relationship to methods of using business simulation. *Simulation & Gaming, 42*(1), 64–84.

Tannenbaum, S. I., & Cerasoli, C. P. (2013). Do team and individual debrief enhance performance? A meta-analysis. *Human Factors, 55*(1), 231–245.

Tebbens, D., & Thompson, K. (2009). Priority shifting and the dynamics of managing eradicable infectious diseases. *Management Science, 55*, 650–663. https://doi.org/10.1287/mnsc.1080.0965.

Thatcher, C., & Robinson, J. (1985). *An introduction to games and simulations in education. Simulations*. Hants: Solent.

Tversky, A., & Kahneman, D. (1974). Judgment under uncertainty: Heuristics and biases. *Science, New Series, 185*(4157), 1124–1131.

7. System Dynamics and ILEs

Alessi, A., & Kopainsky, B. (2015). System dynamics and simulation-gaming: Overview. *Simulation and Gaming, 48*(2–3), 223–229.

Forrester, J. W. (1961). *Industrial dynamics*. Cambridge, MA: Productivity Press.

Homer, J. B., & Hirsch, G. B. (2006). System dynamics modeling for public health: Background and opportunities. *American Journal of Public Health, 96*(3), 452–458.

Meadows, D. L., Fiddaman, T., & Shannon, D. (1993). *Fish Banks, Ltd.: A micro-computer assisted group simulation that teaches principles of sustainable management of renewable natural resources* (3rd ed.). Durham, NH: Laboratory for Interactive Learning, Hood House, University of New Hampshire.

Plate, R. (2010). Assessing individuals' understanding of nonlinear causal structures in complex systems. *System Dynamics Review, 28*(1), 19–33.

Qudrat-Ullah, H. (2007). Debriefing can reduce misperceptions of feedback hypothesis: An empirical study. *Simulations and Gaming, 38*(3), 382–397.

Qudrat-Ullah, H. (2010). Perceptions of the effectiveness of system dynamics-based interactive learning environments: An empirical study. *Computers and Education, 55*, 1277–1286.

Qudrat-Ullah, H. (2014). Yes we can: Improving performance in dynamic tasks. *Decision Support Systems, 61*, 23–33.

Qudrat-Ullah, H., & Kayal, A. (2018). How to improve learners' (mis) understanding of CO2 accumulations through the use of human-facilitated interactive learning environments? *Journal of Cleaner Production, 184*, 188–197.

Qudrat-Ullah, H., Saleh, M. M., & Bahaa, E. A. (1997). Fish Bank ILE: An interactive learning laboratory to improve understanding of 'The Tragedy of Commons'; a common behavior of complex dynamic systems. *Proceedings of 15th International System Dynamics Conference*, Istanbul, Turkey.

Qudrat-Ullah, H., & Tsasis, P. (Eds.). (2017). *Innovative healthcare systems for the 21st century*. New York: Springer.

Rouwette, A., Großler, A., & Vennix, M. (2004). Exploring influencing factors on rationality: A literature review of dynamic decision-making studies in system dynamics. *Systems Research and Behavioral Science, 21*, 351–370.

Spector, J. M. (2000). System dynamics and interactive learning environments: Lessons learned and implications for the future. *Simulation and Gaming, 31*(4), 528–535.

Sterman, J. D. (1989). Modeling managerial behavior: Misperceptions of feedback in a dynamic decision making experiment. *Management Science, 35*, 321–339.

Sterman, J. D. (1994). Learning in and around complex systems. *System Dynamics Review, 10*, 291–323.

Sterman, J. D. (2000). *Business dynamics: Systems thinking and modeling for a complex world*. New York: McGraw-Hill.

Sterman, J. D. (1992). Teaching takes off – Flight simulators for management education, *OR/MS Today*, pp. 40–44.

Sterman, J. D., & Dogan, G. (2015). I'm not hoarding, I'm just stocking up before the hoarders get there. Behavioral causes of phantom ordering in supply chains. *Journal of Operations Management, 39–40*, 6–22.

Sterman, J. D., & Sweeney, B. (2007). Understanding public complacency about climate change: Adults' mental models of climate change violate conservation of matter. *Climatic Change, 80*(3–4), 213–238.

Sterman, J., Franck, T., Fiddaman, T., Jones, A., McCauley, S., Rice, P., Sawin, E., Siegel, L., & Rooney-Varga, J. (2014). World climate: A role-play simulation of climate negotiations. *Simulation & Gaming*, 1–35.

8. Designs Principle for Learning in and with Simulations

Klabbers, G. (2000). Gaming and simulation: Principles of a science of design. *Simulation and Gaming, 34*(4), 569–591.

Kriz, W. C. (2003). Creating effective learning environments and learning organizations through gaming simulation design. *Simulation and Gaming, 34*(4), 495–511.

Qudrat-Ullah, H., Saleh, M. M., & Bahaa, E. A. (1997). Fish Bank ILE: An interactive learning laboratory to improve understanding of 'The Tragedy of Commons'; a common behavior of complex dynamic systems. *Proceedings of 15th International System Dynamics Conference*, Istanbul, Turkey.

Schön, D. (1984). *The reflective practitioner*. New York: Basic Books.

Tversky, A., & Kahneman, D. (1974). Judgment under uncertainty: Heuristics and biases. *Science, New Series, 185*(4157), 1124–1131.

9. Healthcare and Its Dynamics

Krentza, H., & Gill, M. (2010). The five-year impact of an evolving global epidemic, changing migration patterns, and policy changes in a regional Canadian HIV population. *Health Policy (Amsterdam), 90*(17): 296–302.

Montaner, J., Lima, V., Barrios, R., Yip, B., Wood, E., Kerr, T., & Kendall, P. (2010). Association of highly active antiretroviral therapy coverage, population viral load, and yearly new HIV diagnoses in British Columbia, Canada: A population-based study. www.thelancet.com, *376*, 532-539.

Qudrat-Ullah, H., & Tsasis, P. (Eds.). (2017). *Innovative Healthcare Systems for the 21st Century*. ISBN 978–3–319-55773-1.

Roberts, C. A., & Dangerfield, B. C. (1992). Estimating the parameters of an AIDS spread model using optimization software: Results for two countries compared. In J. A. M. Vennix, J. Faber, W. J. Scheper, & C. A. T. Takkenberg (Eds.), *System dynamics 1992* (pp. 605–617). Cambridge, MA: System Dynamics Society.

Robertson, B., Schumacher, L., Gosman, G., Kanfer, R., Kelley, M., & DeVita, M. (2009). Simulation-based crisis team training for multidisciplinary obstetric providers. *Simulation in Healthcare, 4*, 77–83.

Index